The ANATOMY of CANALS

The Mania Years

The ANATOMY of CANALS

The Mania Years

Anthony Burton & Derek Pratt

TEMPUS

First published 2002

PUBLISHED IN THE UNITED KINGDOM BY:
Tempus Publishing Ltd
The Mill, Brimscombe Port
Stroud, Gloucestershire GL5 2QG
www.tempus-publishing.com

PUBLISHED IN THE UNITED STATES OF AMERICA BY:
Tempus Publishing Inc.
2 Cumberland Street
Charleston, SC 29401
www.tempuspublishing.com

© Anthony Burton & Derek Pratt, 2002

The right of Anthony Burton & Derek Pratt to be identified as the Authors of this work has been asserted by them in accordance with the Copyrights, Designs and Patents Act 1988.

All rights reserved. No part of this book may be reprinted or reproduced or utilised in any form or by any electronic, mechanical or other means, now known or hereafter invented, including photocopying and recording, or in any information storage or retrieval system, without the permission in writing from the Publishers.

British Library Cataloguing in Publication Data.
A catalogue record for this book is available from the British Library.

ISBN 0 7524 2385 1

Typesetting and origination by Tempus Publishing.
PRINTED AND BOUND IN GREAT BRITAIN.

Contents

Introduction		6
1.	A New Start	7
2.	The Grand Junction	24
3.	Towards a Grand Union	45
4.	Waterways to Wales	60
5.	South Wales	75
6.	Southern England	82
7.	Manchester and the North	95
8.	The Ship Canals	116
9.	Lost and Found	130
Gazetteer		149
Index		152

Introduction

Volume 1 dealt with the early years of canal construction, the years dominated by the work and ideas of James Brindley. It is thanks to him that a standard was set through much of the system, which involved building locks to take vessels that were roughly 70ft long and 7ft beam, the familiar narrow boats which have become synonymous with canal travel in England, if not in Scotland and Wales. It is arguably unfortunate that such modest dimensions were accepted, though it has to be remembered that they did not seem quite so modest at the time. The most efficient motive power available was the horse, and a single horse hauling a boat was moving perhaps fifteen times as much as a horse and cart on even the best road – and best roads were not very common. By the time the second wave of canal construction began, the web spreading out from the English Midlands was so well established that attempts to introduce grander canals was mainly limited to areas which had no connection with the main system. The narrow boat continued to rule for as long as commercial carrying continued.

Engineers wisely accepted that, short of rebuilding the old at great expense, there was little to be done other than accept the world as they found it. This is not to say, however, that they were in any way obliged to follow the pioneers in other matters. Where Brindley had, as far as possible, avoided problems and major engineering works by following natural contours, even at the expense of adding many miles to a route, the new generation was prepared to tackle obstacles head on and evolved a new vocabulary of civil engineering. As readers will discover in the following pages, the Age of Brindley was about to give way to the Age of Jessop. It was an age that was to give the canal world some of its most memorable and dramatic features.

1. A New Start

Canal building that had stuttered to a standstill in the 1780s, with just two or three new projects being pushed forward, was revitalised in the 1790s. Suddenly there were new schemes being promoted with such astonishing rapidity, with no fewer than twenty-one Canal Acts being passed in 1793 alone, that the early part of the decade became known as the time of Canal Mania. It was not, however, just an enthusiasm for canals that set the period apart. A new generation of engineers began to find new ways of solving old problems and to use the new technologies of the Industrial Revolution.

As good a place to start as any is a little canal, actually begun in 1788, but soon to be developed into a mini-network serving the booming industries of Shropshire. The story begins with one of the great iron masters of the area, William Reynolds of Ketley. There was a valuable source of coal and ironstone about a mile and a half from the works, but over 70ft above it. He set about building a canal, but it was clearly ludicrous to build the half dozen or more locks that would have been needed to overcome the differences in level. As this was his very own private canal, Reynolds decided to experiment with something quite new to the world of canals. Out went the locks and in came the inclined plane. It consisted of a double track of iron rails on which were mounted specially constructed frames, with large wheels at the front and small at the back so that the platform remained level. A boat would enter a lock at the top of the incline, and water was then let out into a side pond, allowing the boat to settle down onto the frame for its journey. Meanwhile a second boat at the bottom of the slope on the adjoining track was floated onto its frame. The two were connected by ropes wound round a drum. The top boat with its load from the mines was always heavier than the bottom boat so that, as the brake was released on the drum, the weight of the descending boat hauled up the empty or partially filled vessel. So a 73ft vertical drop was accomplished in one action instead of half a dozen and with no loss of water – the water from the side pond was simply fed back into the lock for the next journey. To make the system workable, Reynolds designed 'tub boats', just 20ft long by 6ft 4in wide which were used singly on the incline, but could be hauled in trains on the level canal.

There is little left to see here, but this was just a beginning. One of those who came to see and admire, and who was shortly to become involved in canal construction in the area, was a young Scottish engineer, Thomas Telford, who wrote:

> *As soon as the plan of ascending and descending by means of an inclined plane was fairly understood, every person was convinced that its principle was very applicable to the situation of the ground which lay between the Oaken Gates and the River Severn, and that this invention alone would obviate the difficulties which before had been considered insurmountable. Under this impression a subscription was entered into, and an Act of Parliament was obtained for the Shropshire Canal.*

This canal, the Shropshire – not to be confused with the Shropshire Union – took a line from the Ketley Canal towards the Severn near Coalbrookdale, and then turned east along the river valley past Madeley, for a final plunge to river level at Coalport. There was one extraordinary feature connecting the canal to the Coalbrookdale iron works. At Brierley Hill, a tunnel was dug at the lower level and laid with rails. Boats arrived at a basin at the upper level, where raw material was loaded into special crates and swung over by a crane to a shaft leading down to the tunnel. Here the crates were dropped down onto bogies and wheeled out of the tunnel – an early form of containerisation. The system did not last. By 1794, an incline had been built to take over the work.

Little now remains of the eleven-mile-long canal, but part of the upper section above the Severn now runs through the Blists Hill open air museum site, while the lower section can be seen by the Severn at Coalport. In between these is the splendidly restored Hay Incline, just over 1,000ft long, with a vertical fall of 207ft. It is very similar to the Ketley Incline in operation, except that there was also a steam engine that could be used for hauling up loaded tub boats from the lower level when required. The system remained in use until the 1890s.

The first iron trough aqueduct, built across the River Tern at Longdon. The abutments of the original, conventional aqueduct planned for the site can be clearly seen.

The other main features of interest can be seen at Blists Hill. There are three stop locks which, when closed, allow sections of the canal to be drained through a valve. These are simple affairs, where the canal narrows in and grooves in the side walls allow planks to be dropped in. This was very necessary, for the canal has a precarious situation on a steep hillside. Not long after the opening in 1793, a landslip closed it again.

This was an area where, in the eighteenth century, great projects were formed. Abraham Darby first used coke instead of coal to make cast iron from a blast furnace, and he was soon to show how valuable a material it was by building the famous Iron Bridge across the Severn that was to give its name to the town. These are only two, but probably the most important, examples of the way in which the region was a forcing ground for innovation. It seems inevitable that canals built in this area were not going to conform to generally accepted patterns. In 1793, a further connection was made when an Act was passed for construction of the Shrewsbury Canal. It was to run from the Severn to the Castle Foregate basin in the heart of Shrewsbury and run on to join the Shropshire. The canal was twelve miles long with one very substantial inclined plane, the Trench Incline, and what was, for its time, a generously constructed tunnel. It was 970 yards long and 10ft wide, with a 3ft wide towing path, cantilevered out over the water on brackets – an idea which Telford was to borrow when building the second Harecastle Tunnel on the Trent & Mersey Canal (Vol.1). The other notable feature was an aqueduct across the river at Longdon-on-Tern. As the River Tern was not navigable, all that was required was a simple, conventional low structure, and that was exactly what the engineer Josiah Clowes designed. Visit the site today and you can see how the canal approaches the river bank over stone arches. Two things then happened – flooding destroyed the early work on the river crossing and Clowes died. His place was taken by Thomas Telford who had just begun work under William Jessop on the Ellesmere Canal (p.61). Here was a young man, full of ideas and eager to make his way in the world, who was fortunate enough to find himself working on a canal, one of whose principal promoters was that innovative iron master, William Reynolds.

Telford had already been thinking about using cast iron for aqueducts, and in his papers he left a drawing of an alarmingly unstable looking structure, carried on spidery iron trestle piers. This was doodling with ideas, for he was in no position to put his idea into practice at that time. Now, however, he was in charge of a canal as chief engineer, and had access to a man who understood construction in iron as well as anyone in the world. Together they decided to scrap Clowes' plan for a conventional aqueduct, and to carry their canal across the Tern in an iron trough. Although the canal has gone, this historic structure survives. The trough rests on three triangulated supports set on stone foundations. The 187ft long trough itself is made up of twenty-six separate cast iron sections, bolted together, with a separate towpath outside the trough. Because part of Clowes' original still stands, it is possible to make direct comparisons. Where the former is very much in the style of Brindley, with massive abutments to support the masonry trough with its heavy puddled clay lining, the iron aqueduct seems to float across on the flimsiest of supports. There is very little doubt that Longdon-on-Tern was used as a testing ground for a new idea, which was soon to be developed in altogether grander style in what is one of the finest, and most famous, canal structures in Britain, as we shall see later. Of the rest of the canal there is little to be said. The line can still be traced through the centre of Shrewsbury, confirmed by the three-storey Canal Tavern, but there is little else of significance. We are lucky that the two great innovations that mark this little network of canals, the inclined plane and the iron aqueduct, still survive.

The imposing offices of the Worcester and Birmingham Canal Company at King's Norton where the canal makes a junction with the Stratford.

These canals brought improved transport to a burgeoning industrial region, but the great centre of early canal development, Birmingham, had not been forgotten. Here canals had proved their worth, if anything rather too well. There was growing resentment against the meanders and delays of the old system. The first link between Birmingham and the Severn had been made through the Birmingham Canal, wobbling its way across the plateau, then swooping down the long flight of locks at Wolverhampton, before continuing on an equally wayward route via the Staffs & Worcester to Stourport. Before the improvements that were made to the Birmingham Canal in the nineteenth century, this meant a journey of forty-seven miles. Even then, for vessels heading south, the problems were not over, for there was a notorious ten mile stretch between Stourport and Worcester, where shallows and sandbanks made river navigation decidedly difficult. A new canal company was formed, the Worcester & Birmingham, offering a journey of just thirty miles instead of fifty-seven between those two places, and the Act was duly passed in 1791. But those sort of cuts in journey distances are not obtained just by hunting for a better line: there was a high price to pay in engineering works – far higher than anyone thought when work began.

The initial survey was carried out by Josiah Clowes, together with John Snape. The chosen line may have been direct, but it involved five tunnels, four of them comparatively modest, but the fifth, variously known as Wast Hill, West Hill and King's Norton was an immense 2,726 yards. Equally alarming was the engineers' decision that seventy-six locks would be needed to overcome the difference in levels between Birmingham and the Severn. And, as if that was not enough, the plans called for a barge canal, with wide locks and, of course, wide tunnels. This was not quite as perverse as it might seem. Both the Staffs & Worcester and the Birmingham Canal Companies had opposed the Worcester & Birmingham, on the grounds that it would

The stop lock marks the end of the junction between the Birmingham Canal and the Worcester & Birmingham, and is crossed by a typical cast-iron bridge.

take away their trade – which was exactly what it was designed to do. They lost the main battle, but the Birmingham achieved one telling victory. They may have been faced with traffic draining away down the Worcester & Birmingham, but they had no intention of seeing their precious water supplies disappearing as well. A clause in the Act decreed that the Worcester & Birmingham could approach no nearer than seven feet to the Birmingham. The result was a gap in communication; the Worcester Bar. So, if no boats were going to be allowed to cross the Bar, then there was a lot to be said in favour of having a canal that could take barges from the river, without trans-shipment. It was such a sound economic argument that no one seemed unduly worried by the practical difficulties, such as keeping the canal in water – no fewer than ten reservoirs were suggested, which represented another colossal expenditure.

The Worcester & Birmingham can be taken, in many ways, as typical of the new mood not just among canal bulders, but in the country as a whole. The economic slump that had accompanied the disastrous war in America of 1775-1783 – disastrous to the British, that is – was over and trade was booming, industry thriving. It seemed that anything was possible: money was no object, all physical barriers could be overcome. The history of the actual construction shows how blind optimism had conquered common sense. The initial enthusiasm had been immense: the press at the time reported that Parliament had received a petition in favour of the Bill with 4,058 signatures, that 'measured in length above fourteen yards'. When work began on the northern end, it was still blithely assumed that a barge canal could be built with the huge lockage, tunnels and reservoirs for £180,000, but there was provision for raising a further £70,000 if needed. The first fourteen miles from Birmingham were lock-free, and followed a reasonably direct line, apart from a pronounced wiggle round the hills at Alvechurch. The first obstacle to be overcome was the ridge at Edgbaston which

Tardebigge tunnel, one of five on the Worcester & Birmingham, has a plain, unadorned portal.

was pierced by a short tunnel. It was built to generous dimensions, with a width of 16ft and a height of 13ft 3in above water level, and supplied with a towpath. By the time the far more formidable obstacle of Wast Hill had been reached, sitting firmly across the line of the canal, a certain caution had set in. The width had to remain the same, but the height was reduced to 10ft 9in, and a towpath was now considered an unaffordable luxury. Two more tunnels of the same dimensions brought the canal to the top of the daunting hill at Tardebigge. By now, without a single lock having been built, the company had spent all the money and had to return to Parliament for authorisation to raise more. In 1798, they were permitted to sell extra shares to the value of £150,000, but when they failed to find buyers, the company were back again in 1803 for permission to raise the missing £50,000 from existing shareholders. The days of easy optimism had ended. All work stopped, and no one seemed quite sure what to do. One thing at least was clear – the original plan for a wide-locked barge canal had been torn up.

The 30 lock flight at Tardebigge is the longest in Britain, but because they are never all in sight at once, they never appear very daunting.

A new engineer, John Woodhouse, was called in and he came up with a novel suggestion. He would reduce the number of locks from seventy-six to twelve, by replacing the other fifty-four by vertical lifts. An experimental lift was built at the top of Tardebigge Hill, but the proprietors were unconvinced, and turned instead to a man with a solid reputation, John Rennie. He offered a better plan than Clowes', if not so radical as Woodhouse's. He proposed a total of fifty-eight locks, thirty of which would be grouped together into what was, and still is, Britain's longest flight at Tardebigge, with a total fall of 217ft. The top lock is unusually deep with a drop of 14ft, not from any choice of Rennie's but because it occupies the site of the experimental lift. Once Rennie's plans had been accepted, work went steadily forward, but not without another visit to Parliament to raise a further £208,000 in 1808. By the time the canal was opened in 1815, the work that had been estimated at £180,000 for a broad canal had been completed as a narrow canal for £552,000. These were, indeed, the years of Canal Mania. In spite of this, however, the canal is a fine one and was to prove very profitable for many years, helped by a decision to open up the Worcester Bar after all. But those who come this way today will be very conscious of the main line railway from Gloucester to Birmingham, which is never far away and is sometimes a very close companion indeed. Boatmen making their way through Edgbaston, their boat being pulled gently along, usually on this canal for some reason by a pair of donkeys, must have been galled by the sight of the steam engine chuffing past them, the driver literally looking down on their leisurely progress.

We can now look at the canal in rather more detail, starting at the Worcester end. The question of where to make the junction with the river was determined partly by the natural landscape and partly by the man-made. Worcester has been a walled city in the past, with the cathedral as its focal point. There was never any question of forcing a way through the tightly packed centre, so a decision had to be taken based on where land was available for a basin, and whether it was easier to go north or south of the built-up area. The southern route proved the easier, and a junction was made upstream of the Diglis river locks. As at Stourport, connection between river and canal was via two substantial barge locks, 76ft long by 18ft 6in which led to a double basin just off the main line of the canal, where cargoes could be exchanged between river craft and narrow boats. Unlike Stourport, where the canal had arrived at an almost deserted section of river bank, the Diglis basins lie at the edge of the city, so there was no need for any great flurry of building activity to provide houses, hotels and offices for merchants. The basin has its warehouses, toll house and dry dock, but all these are strictly functional, the minimum requirements for the trade of the canal. It was here that general cargoes were stored – grain in large quantities and, at the luxury end of the market, wine sent up by importers in Bristol. Once the basins are left behind, broad locks give way to narrow.

It was inevitable that having been pushed out to the back end of the city, the canal would get only the occasional glimpse of the city at its grandest, though it does enjoy a view of one of the finest of the old buildings, the Commandery, a Tudor house built on the site of the ancient St Wulstan's Hospital. Beyond that, what there is of interest comes largely from the canal itself and its later neighbours. The waterway is crossed by a tall railway bridge linking the two Worcester stations, Shrub Hill and Foregate. It is carried across the canal on a single high arch, but where it crosses the adjoining road a lower, round-headed arch does the job. To reduce the material needed, without affecting the strength of the arch, a circular hole has been left in the spandrel. The effect is not unlike a giant exclamation mark. Rather more delicately, a canal bridge rises in a gentle curve to carry the towpath over the entrance to Lowesmoor basin. The elegance of the bridge is simply a response to a practical problem. The gentle slope enables the pull on the towrope to be maintained, where a steeper slope would tend to cause a lift on the bows and an increased load for the horse or donkeys. Lowesmoor basin was originally used for bulk cargoes, mainly timber and coal.

Once out from under the railway arch, the canal begins to turn away towards the countryside and the steady procession of locks that will culminate in the final steep climb at Tardebigge. The locks themselves are conventional, with single gates at the upper end and double at the lower, but several have an unusual type of bridge. A number of canal engineers came up with the idea of split bridges at locks, to allow the tow rope to pass straight through the middle. Here the solution is even simpler. The bridge is a simple cantilevered platform, firmly supported on one side of the lock only, rather like a diving board. The unattached end stands slightly proud of the lock wall leaving a gap for the rope. The gap will automatically be closed by the weight of anyone actually walking on the bridge. This simple idea did not find many imitators, though it was used on the improved Birmingham main line, begun in 1827. The lock cottages, in contrast, hold no surprises, being once again plain, with regular facades, with a little variation supplied by the segmental arches above the doors and ground floor windows. For those who have ever wondered about the interiors of lock cottages, the one halfway up Tardebigge has been fully restored in traditional style by the Landmark Trust and can be rented for holidays.

There are a number of key canal points on the route from Worcester to Tardebigge. The first is Hanbury Wharf, where the canal is crossed by the B4090, that runs along the line of the ancient Salt Way. This is not unimportant. Salt from Droitwich had been a valuable commodity since Roman times, and the Droitwich Canal had already been built specifically to profit from the trade, taking boats down to the Severn. By the mid-nineteenth century, much of this valuable cargo was going by rail, so to combat the loss a new canal, the Droitwich Junction, was begun in 1852. It was just 1 ½ miles long, linking Droitwich basin in the town centre with seven locks, of which one was built as a barge lock, as used on the original Droitwich Canal, and the rest were narrow. The barge lock was also used to control the flow of water from the River Salwarpe, which was incorporated into the new canal for 160 yards. As a result, Hanbury Wharf became an important centre of development with cottages, an unusual house for the manager of the wharf and adjoining yard, serving the two canals and the Salt Way. The house has a curved end to accommodate it to an awkwardly angled site. Next door is a former canal pub, which once had its own brew house. Salt appears again a little further north at Stoke Works, where brine pumping began in 1828, production eventually matching that of Droitwich. But in spite of the efforts to use the canals for these conveniently situated sites, it proved impossible to prevent the onward march of the railways. Things became so bad in the 1860s that a proposal was put forward to close both Droitwich Canals and the Worcester & Birmingham and use the land for railways. That came to nothing, but the Droitwich canals finally succumbed in 1939. The Worcester & Birmingham lives on and there is nothing along the line of the canal to recall those traumatic times. A little way beyond the brine works is Stoke Wharf where the plain simple brick buildings – cottages, warehouse and pub seem untouched by time.

The Tardebigge locks dominate the next section with, not unreasonably, canalside pubs along the way for the hard worked boaters. Even more important than keeping boat crews topped up was the need to keep the canal itself supplied with liquid. The first of the reservoirs puts in an appearance half way up the flight and from here water was pumped to the summit, and the remains of an engine house can be seen beside the top lock. Now there is a real flurry of activity, with a side arm leading to the other obvious necessity for keeping this long flight of locks open; the maintenance yard. Then comes the first of the tunnels, 580 yards long and hacked out of the sandstone, soon followed by Shortwood Tunnel, of similar dimensions to Tardebigge, but notably drippier. The canal struggles to keep a level as it winds its way through the undulating ground around Alvechurch, while a little aqueduct was needed to carry the canal over a hollow, at the bottom of which a lane leads to Alvechurch station. This section seems altogether more hesitant than the direct, confident line taken by the section engineered under Rennie. Now the canal is literally overshadowed by its two great competitors: first the railway branch line from Redditch to the main Worcecster-Birmingham line, followed by the latest addition to the transport scene, the M42.

The Worcester & Birmingham was always a thirsty canal, and in the 1830s two extra reservoirs were built, the Upper and Lower Bittell, supplying the canal through a half-mile-long

Kingswood Junction where the Stratford Canal meets an arm of the Grand Union. The lock on the right is crossed by a split bridge, allowing the tow rope to pass through.

feeder from the upper reservoir. These man-made lakes have taken on a very natural air and what were once no doubt seen as unwelcome intrusions into the natural landscape are now highly valued wildlife habitats. They are there to be enjoyed as well as providing an essential service. It is as well to enjoy this pleasant scene, for those who come this way by boat will soon be plunging into the long King's Norton Tunnel, and when they emerge at the far end it will be to a very different environment. As already mentioned, when the tunnel was constructed it was built to accommodate wide boats, and it was only later that the decision was made to reduce the lock size. Had it been considered as a narrow canal from the first, the tunnel might have been built to less generous dimensions, and would have been a considerable bottleneck, so something good came out of the early period of unwarranted optimism. As it was, from very early on, traffic movement was speeded up by the use of steam tugs. With the advent of motor-powered long boats, the tunnel presented no problems – though the author recalls an alarming encounter here some years ago when he found himself confronted by a flotilla of ordinary wooden doors making a stately progress through the dark. The tunnel does possess one unusual feature; a line of iron brackets in the roof that once held telegraph lines.

At the far end the countryside has been left behind, as the canal reaches the outlying suburbs of Birmingham, with six miles still to go. This does not mean that the canal is now surrounded by buildings for the rest of the way, and it certainly does not mean that it now lacks interest. A little way beyond the point where another feeder from yet another reservoir joins the canal, a junction is met at King's Norton. This was another important link for the Worcester & Birmingham – the canal to Stratford and the River Avon. It is marked by the company offices at the junction house, which has the air of a grand, and perhaps slightly pompous, Victorian villa. The main entrance is under a pillared porch that stands at the front of a substantial two-storey bay, and the door itself is unusually decorative, with hexagonal panels. Stone surrounds provide a weighty contrast to the dull red brick of the walls, and the hipped roof is half hidden behind a parapet. The total effect does not provide the elegant simplicity of the lock cottages, nor is the building quite big enough to carry the elaboration heaped upon it.

The next real interest comes at Bournville. Cadbury's established their works here in 1879 and built extensive wharves, covered by canopies. The whole site was developed with a model village for the work force, with villas and gardens that were to set a standard for suburbia at its best and leafiest. The canal was vital to the whole enterprise. The main ingredients, evaporated milk and chocolate crumble, were delivered by boat from Frampton on the Gloucester & Sharpness Canal, Knighton on the Shropshire Union and from London docks. Cadbury's owned a fleet of boats, but other companies, notably Severn & Canal and Fellows, Morton & Clayton were also involved. The trade continued right on into the 1960s under British Waterways, and one boatman who became a regular on the run earned himself the name of 'Chocolate Charlie' Atkins.

There was a junction just beyond this point, where a new canal had been constructed to join the Dudley – and given the rather unimaginative name of Dudley No.2 Canal. This has now been filled in, and the Worcester & Birmingham continues on past the old entrance towards the prominent tower, reminiscent of an Italian campanile, of Birmingham University. For a time the city is forgotten, as the canal heads down a leafy cutting, with the railway alongside. There is a brief interruption for the short Edgbaston Tunnel before the canal reaches the end of the line. When the Worcester & Birmingham was finally completed, the Birmingham relented and agreed to a junction of the two waterways. The Worcester Bar was

Looking like a giant pepper pot, this ventilation shaft from Gosty tunnel on Dudley No.2 Canal rears up in a suburban garden.

replaced by a stop lock that prevented the movement of water just as effectively as the Bar had done. So the canal reached Birmingham, and the terminus was constructed around Gas Street basin. There was a time when this was one of Birmingham's best kept secrets, a square of water hidden from view from the street by a high brick wall. Only those in the know recognised a wooden hatch, painted red, which signified a fire point: the hatch would be opened and a hose dropped in to pump canal water. The rest of the basin was surrounded by old, brick warehouses. Of the original buildings, little now remains, apart from the toll office by the stop lock. Everywhere else the buildings are new, part of the huge rebuilding of central Birmingham which has taken the canal to its heart as an attraction, rather than a rubbish dump for supermarket trolleys. A new, glass-fronted pub looks down on an array of traditional narrow boats moored in the basin. These at least are reminders that this was once a busy commercial centre, but the old sense of intimacy and privacy has gone for ever. The final connection to the main Birmingham system from the stop lock is via the short tunnel under Broad Street.

Turning back to King's Norton brings us to the Stratford-upon-Avon Canal, whose existence was totally dependent on the building of the Worcester & Birmingham. It seems, on the face of it, to have been a sensible sort of canal, providing a useful connection to a navigable river and, although the gradient was steep, it was only twenty-five miles long. In fact, the idea of creating an inland port at Stratford was not new. Today the only industry we associate with the canal is the Shakespeare industry. There was, in fact, a tradition of textile and leather manufacture – Shakespeare's own father was a glove maker. In 1677, a gentleman named Andrew Yarranton, who had been much impressed by what he had seen of water transport in mainland Europe, and especially Holland, proposed a scheme for two new industrial centres in the town, to be based on waterways – New Brunswick to brew beer, New Haarlem to make linen. Nothing happened for over a century, until 1775 when a canal plan was drawn up, and dropped again until 1793, when the Act for the present canal was passed. Once again Josiah Clowes was called in as chief engineer, a man fated it seems to start much and finish nothing. The rate of progress on the Stratford makes that on the Worcester & Birmingham seem downright hasty. After the initial Act there were to be six further Acts to raise money to complete the work, the last in 1821, more than a quarter of a century after work began. During the period of construction, the canal advanced at a rate of just about one mile a year.

It is necessary to have a brief glance at this lengthy history, as the frequent pauses, interruptions and rethinks meant that the character of the canal changed over time as different engineers were called in to try and complete the work. At first things went quite well, and the canal was completed from King's Norton to Hockley Heath, a distance of some eleven miles. There was even a brave attempt to mechanise construction with the introduction of a horse-drawn cutter, but it proved a complete failure. So, it was back to navvies, picks and shovels. Then the money ran out for the first time. Work restarted in 1800 under Clowes' former assistant Samuel Porter. By now, a new canal had appeared in the area, the important Warwick & Birmingham, part of a proposed new, improved route from the Midlands to London. It was decided to make a junction with the Stratford at Kingswood, and once again things jogged along nicely. Kingswood was reached, and now at least the Stratford had connections at both ends, and the northern part was open for trade. It was, however, still thirteen miles from Stratford itself. The southern section was completed by a third engineer, William James, a man who was soon to lose his enthusiasm for canals and develop a passion for railways. He saw it through to completion, but his was not to be the last name in the long history of establishing navigation along this modest waterway, for it was destined to fall into sad disuse and be rescued as the first major canal restoration scheme in Britain. William James' successor, more than a century later was hustling, bustling David Hutchings who with a motley work force that encompassed everyone from Boy Scouts to prisoners from Winson Green, saw the canal enjoy its second opening in 1964.

The obvious starting point is King's Norton Junction, with its typically sinuous junction bridge, adapted to take the two towpaths. It is a complex medley of curves, all designed to ease gradients for the draught animals and to avoid sharp edges meeting towropes. The most striking feature in the middle of what is now a somewhat glum area of dull housing estates and duller industrial units is the guillotine stop lock in which the gates are held in iron frames and move vertically, counterbalanced by weights suspended from chains passing over a complex pulley system. Now that all the canals are part of the same system, sharing water is no longer a problem and the gates remain permanently raised and, one would imagine, all but unmoveable. This is a canal where consistency is a rarity. After the conventional bridges at the start, the first moveable bridge appears, swinging horizontally. This is soon followed by the 352 yard long Brandwood Tunnel, its portals graced by the bust of Stratford's most famous resident, the bard himself. As far as one knows the nearest Shakespeare ever got to a mention of inland waterways was a description of Cleopatra's barge with its golden poop, silver oars and purple sails 'so perfumed that the winds were lovesick with them'. The horse-drawn narrow boat could hardly compete, and though one can imagine some cargoes which might induce queasiness, it would hardly be lovesickness. There is no towpath in the tunnel, but boats could be hauled along by handrails set into the walls. It was built to the double width of 16ft to allow boats to pass each other.

Apart from the tunnel there were no major problems to be overcome as Clowes followed his preferred method, reminiscent of the Brindley Age when he learned his trade, of following the natural contours as far as possible, but he was willing to be rather more daring, if need be, to keep his level and avoid lock building. He built a small aqueduct across the River Cole and the adjoining road, but faced a more daunting task when faced with the low hills bordering the valley of the Blythe. Here he used a technique that was to characterise the second age of canal construction; 'cut and fill'. Deep cuttings were sliced through the rising ground, and the spoil was then carried away and used to build up an embankment across the valley. The cuttings create private, secluded worlds in which the boater has little idea whether the surrounding country is urban or rural. The banks, by contrast, offer panoramic views. Bridges are generally on a very generous scale, even when carrying no more than a farm track. There is one more moveable bridge on this section, but this time a lift bridge not a swing bridge. It was lifted by hauling down on a chain attached to the end of an overhead balance beam. It has now been replaced by a steel bridge, worked by a windlass. There was a shop here, known as the Drawbridge Stores, which was at one time a pub, The Boatman's Rest, and now the two have combined in a new pub, The Drawbridge. There was to be one further development on this section, when an Act of 1815 authorised the construction of reservoirs at Earlswood, which became known as the Earlswood Lakes, and became so popular with visitors that when the railway was built between Birmingham and Stratford, the company created a special Lakes Station. During the years when Hockley Heath represented the end of the line, the canal was served by a small basin, approached under a very narrow, but high arched bridge. Today it is a reedy, silted backwater, and the former warehouse has become the Wharf Tavern.

The second phase of construction saw the start of economies, with bridges scaled down from the grandeur of the early days, and it also marked the start of real lock building, eighteen of them down to the new junction at Kingswood. The first four are widely spaced, but after that they come tightly packed in the Lapworth flight. There could have been a water problem created by having very short pounds, but in the lower section the pounds have been widened out to create shallow reservoirs. To avoid the necessity of constantly unfastening the tow rope to get under bridges, cast iron split bridges have been constructed. The name 'split bridge' is really rather misleading, suggesting that a bridge was built and the centre cut away. In fact these are cantilevered platforms, not quite meeting in the centre. They are widely regarded nowadays as being rather charming, but at the time they were built this was all new technology. When the canal age began the first iron bridge had yet to be built, but now engineers were beginning to realise the advantages of cast iron. The parts are created by making a full scale replica in wood, which is then packed with a special sand and the pattern is then removed leaving its 'negative' behind. When molten iron is poured in from the furnace, the pattern is duplicated in metal. This can be done over and over again. As canal bridges are necessarily built to a standard size on any one canal, they can be supplied with standard bridges. The Stratford bridges may have an old-world appeal, but this is the start of mass production.

At Kingswood itself, a short arm leads off to join the Warwick & Birmingham, now part of the Grand Union. At such an important point where, once again, the Stratford paused to recoup its resources, there are, as one would expect, a number of important buildings, including lock cottage and maintenance yard. The latter has no sense of pretension, but is visually very satisfying, its dark brick set off by stone quoins, with prominent keystones above the waggon arches. This northern section of the Stratford remained in use and was never allowed to fall into dereliction, as it was a useful adjunct to the Grand Union, not least for traffic for Bournville. The southern section was the last to be built, and the least needed. A very good case could have been made for ending the canal here, but the builders pushed on to Stratford and the Avon link. There was never a great deal of traffic and it gradually fell into such a state of neglect as to be all but unusable, until rescued by volunteers. But the fact that it served no great industries and visited no major towns, were the factors that led to its near disappearance and were just the things that made it so appealing for the growing holiday trade. It also has a character very much its own, very different from its northern half.

Passing through undulating country, canal building was always going to involve considerable engineering works. There were only thirteen miles left to build, but into that space were crammed a further thirty-six locks and three aqueducts. Money was scarce, so the hunt was on for economies. There was nothing to be done about locks, but cash could be saved on lock cottages. Instead of the familiar, conventional two-storey cottage with pitched roof, the southern Stratford has single-storey cottages with barrel vaulted roofs, ending in neat parapets. It has been suggested that the roof design was chosen because formers were already available for constructing bridge arches, but the profile looks very different. It seems likely that economy of materials is a quite sufficient explanation. As for the story that they were built by navvies who only knew how to build short tunnels – well! It has to be said that tunnelling rarely consists of building walls and a curved roof and then putting a hill on top!

When it came to the aqueducts, James went for the new technology, cast iron. There is a modest beginning at Yarningale, a more imposing version at Wootton Wawen, where the canal crosses the main road from Stratford to Birmingham, but grandest of all is the crossing of a tributary of the Avon, the adjoining road and a later railway at Edstone. James took his inspiration from Longdon-on-Tern, for like that early example, the towpath is carried alongside the trough. The methods of construction can be clearly seen. A series of isolated, narrow piers was built along the line of the canal, like a row of children's building blocks, and the iron trough was sat down on top of them. Lacking the triangulated supports of Longdon, James used inverted triangular trusses beneath the trough to give it rigidity. It is an impressive work in engineering terms, 475ft long, but only reaching a maximum height of 28ft, so that it looks rather heavy and lumpen.

The canal ends at what would originally have been just another canal basin, joined to the river by the usual broad barge lock, for easy interchange. That, however, was before Stratford-upon-Avon became one of the world's leading tourist atractions. Now the commercial basin looks like a lake in an ornamental park, looked down on not by warehouses but a statue of Shakespeare and the theatre where his works are performed. Given this setting it is not difficult to understand the enthusiasm of the modern restorers. The other canal that was built as a response to the arrival of the Worcester & Birmingham on the scene has not been so fortunate.

Windmill End, at the southern end of Netherton tunnel. The prominent chimney belongs to the engine house built in the early 19th century for a colliery pumping engine.

Dudley No.2 Canal originally ran by a devious, but lock-free route from the old Dudley Canal at Parkhead locks for ten miles to join the Worcester & Birmingham at Selly Oak. Its success was due in large measure to its connections with local coal and ironstone mines, but these industries which brought early profits also brought eventual disaster. Far and away the most important engineering features on the canal were the two tunnels. Gosty Hill is a modest 557 yards long, but does have one unusual feature. One of the construction shafts, retained for ventilation, originally rose up into open country, but no longer. Its circular brick tower, topped by a curved iron grill now makes a gargantuan garden ornament in Old Hill. Lapal Tunnel was far longer at 3,795 yards and consequently more vulnerable. Completed in 1798 to the reasonable dimensions of 9ft high by 9ft wide, subsidence reduced the size of the arch above water level, so that in places it was just more than a boat's width. As the tunnel has no towpath, boats were legged through and in that narrow channel water resistance became an important factor. The company tried an ingenious solution in 1841, installing a steam pump which raised the level of the water in the tunnel. By then removing the stop gates, a flow could be created in the appropriate direction, easing the tired limbs of the leggers, and more importantly for the company, reducing waiting time. There was a constant battle to prevent further damage, but in 1917 the battle was lost, and the roof fell in. The tunnel was closed and the whole eastern end of the canal abandoned.

The canal is still open as far as Halesowen, piercing the heartland of the old industrial world of the Black Country. It begins by curling round Netherton Hill, topped by the old village, with the industries – mainly iron works and coal mines – spread out at the foot. Evidence of both activities can be found on the canal. The first part swings past the long abandoned half-mile-long Two Locks Line. It then slices through the shoulder of the hill in a cutting crossed by Highbridge Road. The immense echoing arch has earned the structure the name of Sounding Bridge. It is unusual in that the arch does not spring from abutments based at canal level, but springs direct from the reinforced sides of the cutting. At the end of this section is a feeder from the small Lodge Farm Reservoir, now used by dinghy sailors. The canal then curls up towards the old mining area of Windmill End. Here the problem of pumping water from the deep coal pits was solved by bringing in a big beam engine from Cornwall. In Cornish engines, the engine house is more than a cover to keep out the rain. It has a thick end wall on which the great overhead beam pivots – one end being connected to the piston in the steam cylinder inside the house, and the other half sticking outside and joined to the pump rods disappearing down into the shaft. Even when the engine was sent for scrap, the sturdy engine house remained and is now preserved.

The area has seen many alterations over the years, notably the arrival of a new tunnel built in the 1850s from the Birmingham main line to relieve pressure on the old Dudley Tunnel. So what had been a continuous canal now had an addition at Windmill End Junction. Here one can see a use of iron which is typical of the Birmingham Canal Navigations – its standardised cast iron bridges. These, undated, are from the Toll End Works, and Toll End is the original name of the Horseley Branch Canal. It was here that the Horseley Iron Works established their first furnaces between 1808 and 1809. Most of the cast iron bridges built by that firm are very similar to these, so it is reasonable to assume that they are the product of the same works. The bridges are cast in sections with two flat arches, with integral hand rails, joined together by a central locking plate. The decks are created from smaller iron plates bolted together. These spring from simple abutments of brick or stone.

The canal continues through an old industrial area to Gosty Hill, where there are the

This boat might appear insignificant, but this was among the first iron boats ever built. It was designed for use on the Hay Incline

remains of a covered dock at the tunnel end, once home to the tugs that took over from the leggers. The line now ends at Coombeswood basin, an important canal-rail interchange, for there was a meeting of various rail links here. It was a far busier place then than it is today. All this activity drew on the area around Birmingham, with its rapidly growing economy. By the end of the eighteenthth century it was clear that what Birmingham really needed was a far better route to London, than that offered by the older generation of canals doddering off towards the Thames at Oxford.

2. The Grand Junction

Until the 1790s the connections between Birmingham and London were woefully inadequate. Here, on the one hand, was one of the fastest growing industrial centres in Britain and there, on the other, was the nation's capital and leading port. To travel between the two by water meant a journey down the Birmingham & Fazeley, on to the Coventry, then the Oxford with a last leg down the River Thames. This was a trip of some 230 miles and, to put that in perspective, the modern road journey between the two cities is just 120 miles. It was quite clear that a more direct route would bring huge advantages and immense cost savings. Before work could begin, however, there was to be a good deal of fighting over the route to be taken, and not just on sensible grounds such as engineering difficulty, economics and forecasts of future traffic. The existing canals were anxious to retain as much of the traffic as they could for their own routes, to maximise tolls. So the Oxford canal company were very much in favour of a cross country route to London that would begin at Thrupp, just north of Oxford, which would bring a huge increase in traffic to their canal. The fact that the Oxford itself was a narrow canal, wayward and notoriously slow, was of secondary importance. In the event, the first and most important section of the direct route was to be the appropriately named Grand Junction Canal, authorised in 1793.

The canal was to run from the Thames at Brentford to join the Oxford Canal at Braunston. The initial survey was carried out by James Barnes, a senior engineer in his fifties who had gained his early experience on the Oxford, and he was soon joined by the man who was to carry so much of the burden of the canal mania years, William Jessop. The latter was a solid, hard working, dependable and incorruptibly honest man, who had built up a career steadily over many years. He had first begun work as an apprentice to the great John Smeaton at the age of fourteen, gradually gaining knowledge of canal and harbour construction and land drainage. He rose to be Smeaton's principal assistant and was eventually taking on work in his own right. In 1789 he was appointed chief engineer for the first time on the Cromford Canal in Derbyshire (Vol. 1). He was soon in demand from canal companies up and down the country, but of all the schemes in which he was to become involved this was arguably the most demanding of them all. At first it was intended to run for just ninety-four miles, but that was to be extended by arms to Northampton, Buckingham, Aylesbury and Wendover and, most importantly, Paddington near the heart of London, adding another thirty-one miles to the length. The Slough Arm was not part of the first construction period.

Barnes and Jessop were well aware of the shortcomings of the older narrow canals and planned on a large scale. They hoped, and expected, that the old canals would one day be widened, so they decreed that locks should be built to take river barges, able to carry fifty to seventy tons. To give the narrow boats a role on the new waterway, the locks were built so that two of these vessels, now so generally in use, could fit snugly side by side in the locks and, just as importantly, pass each other in the tunnels. This forethought was to prove of immense advantage in later years. The old canals never were widened as the advent of the railway age sent investors rushing off in new directions, leaving the canals to make the best of what they already had. But with the introduction of the motor boat, the working pair – a motor towing an unpowered butty – the Grand Junction was able to pass the two vessels through locks at the same time. On the older canals, the motor had to go through first and wait for the butty to be manhandled through.

The decision to opt for a broad canal was a bold one. Earlier engineers had shied away from the notion of building wide tunnels of any length, and here Barnes and Jessop were to end up constructing two of them, Braunston (2,049 yards) and Blisworth (3,076 yards). It is a measure of how far canal engineering had moved in a matter of little more than two decades. It was the prospect of having to construct a tunnel through Harecastle Hill that had first caused James Brindley to build the Trent & Mersey Canal to half the width of the earlier Bridgewater. In opting to go back to the wide measure, the new men were not just proposing

A working pair at Home Park Mill: the locks on the Grand Union are wide enough for two narrow boats to sit side by side – motor boat and unpowered butty.

The footbridge marks the junction of the old Grand Junction Canal near Paddington Basin with the Regent's Canal.

to double the amount of material that would have to be excavated, as the work relates to the cross section not the width. They were going to have to dig and blast away four times as much to make the improved route. It was a sensible decision, though given the trouble the tunnels were to cause it might not have seemed so, as the engineers struggled with floods, quicksands and collapses. There were to be other major engineering features – flights of locks, a deep cutting at Tring and aqueducts. It was to take twelve years to complete, but when compared with the forty-six years taken by the other major route, the Leeds & Liverpool, it seems almost indecently hasty.

The basic engineering of the Grand Junction has changed little over the years, but the canal itself was to undergo a major course of improvement in 1929 when it was amalgamated with others to form the Grand Union. Commercial carrying with narrow boats lasted on this canal longer than on any others, and this is invariably reflected in the structures met along the way. Change is seen at its most dramatic at the start in London. For years, canals were regarded as unhealthy backwaters and it is only recently that they have come to be seen as amenities to be cherished, first through the rise of pleasure boating, then as offering centres for new developments – not because the canals were useful, but because they had come to seem quaintly picturesque. Some of these recent developments will be looked at in more detail in the next volume, which brings the canal story up to date. What one can say is that the Paddington basin, which became for a time the terminus of the Grand Junction, has been totally transformed, and it is hard to believe that it once enjoyed a rural setting on the outskirts of London.

Beyond the basin the canal is all but overwhelmed by the later transport systems that were to bring cargo-carrying by water to an end. It scuttles past the end of the immense glass-covered train shed of Brunel's Paddington Station before reaching the concrete flyover of the Westway motorway. It is only then that it really comes into its own, establishing a definite identity. An island marks the junction with the later Regent's Canal as the Grand Junction turns to head west. It was here that the advantages of an urban canal were first recognised, earning the area the inevitable, if not very appropriate name, of Little Venice. There is really very little that is at all Venetian about the scene. Houses stand back from the water in terraces separated from the canal by busy roads: they do not have water lapping round their walls and a gondola at the front door. This is, in fact, a very London scene, of rather elegant early nineteenth-century houses as opposed to the rather debased form of classical terraces that

Berkhamsted locks looking neat and trim are important as marking the official northern limit for barges and other broad-beamed craft on the canal.

typifies so much Victorian building. It is interesting to note that our forebears understood the attraction of water in the urban scene, a lesson we have only recently relearned. The canal, also, has its own points of interest: a handsome iron bridge bears the Paddington coat of arms and at the next bridge is that essential feature of all waterways, the toll house.

The Paddington Arm presented few problems to the engineers who, if not able to follow a completely straight line, were at least able to squirm round the undulations without too many extravagant curves and without having to build a single lock in the whole length of nearly fourteen miles. Since the canal was built, London has spread and been subject to redevelopments which have had their effect on the canal scene. There are memories still of the old life and old buildings. The seemingly fragile lattices of gas holders still survive, though the works that once received their coal by water have long gone. Factories still line the route, and one can choose between aromas, from biscuity McVities to soupy Heinz. Modern tower blocks look down on Victorian churches whose spires once dominated the skyline. The route passes Kensal Green cemetery where Brunel was buried, and Old Oak Common where the giants

The Paddington Arm of the Grand Junction turns away to the left under the elegantly curvaceous bridge at Bull's Bridge Junction.

of the steam age, the Kings and Castles of his GWR, had their sheds, now all rather drab and home to 125s. There is little of the original canal world left, and the first aqueduct to appear is a modern creation. This is, in fact, the second aqueduct on the site. The first followed the creation of the North Circular Road, and when that was widened the present version was constructed. It is only beyond that when Bridge 12 appears that the familiar canal tradition emerges, the humpbacked bridge. It is built to a pattern that will become familiar to anyone travelling this canal. First comes the brick arch, then a pronounced string course, covering a change in the brick laying pattern, above which the brickwork follows the curve of the arch, to be topped by a stone parapet. Such bridges are so common that they scarcely seem to merit a second look, but their unobtrusive simplicity disguises the fact that these are sophisticated structures, built on the curve in plan as well as elevation.

The branch ends at Bull's Bridge, where it meets the original main line which has made its way up from the Thames. So, this description goes back to the other beginning of the canal at Brentford. It is not strictly true to say that the canal joins the Thames, for the first part of the journey is along the River Brent to the Thames locks. The river is tidal here so the locks can only be operated at certain times and they are still very much under the control of a lock-keeper, operating a modern system. Trade lingered on at Brentford, long after it had died elsewhere in the system, when lighters brought cargoes up-river for trans-shipment into narrow boats. Then, as narrow boat traffic dwindled, the goods were taken on by truck. Then the lighters, too, began to disappear. The modern canopied warehouses still speak of a period of optimism, when it was felt that common sense decreed that it was far better to move goods to the outer fringes of London by water than to use clogged city streets. They are still in use, but goods come in and out by road – and mostly they echo hollowly to the sound of a flapping door, sad reminders that an opportunity to do something about pollution and road gridlocks seems to have been lost.

The canal continues up the Brent valley through an area where parkland alternates with new developments, ranging from the high tech of steel frame and tinted glass to the curiosities of the late twentieth century, where the steel frame has an outer skin of brick and is adorned with odd little gables and fronted by pillars supporting nothing. Through parks and past offices, the canal manages to establish its own character. The first locks appear, often with wooden bollards sculpted by countless lines from boats passing through. Bridges give messages of different generations. One is built in a style that will reappear later in the journey, a typical product of the Horseley Iron Works, carrying the words Grand Junction Canal and the date 1820 cast into the ironwork. When, a century later, the Grand Junction was incorporated into the Grand Union, the 'Junction' was chipped away and the word 'Union' crudely painted in. A multi-arched viaduct carries the overground section of the Piccadilly line and the M4 strides over on stumpy, concrete legs. Transport routes all come together at Three Bridges, where the canal hops over the railway and the road crosses above the canal. There are curiosities along the way, such as oddly named Asylum Lock – and the only sign of how the name originated can be found in a bricked-in arch in the wall surrounding the Ealing Hospital. This was once a mental hospital, and boats would pass through to deliver coal. Elsewhere, other arches and openings lead off to arms and docks once busy with trade – a private arm once served the Quaker Oats factory; Adelaide Dock was home to the Murrell's fleet of lighters, among the last regularly to work river and canal. This section ends back at Bull's Bridge, and the typical shapely junction bridge, with its gentle sloped ramp carrying the towpath of the main line across the Paddington Arm, while the Paddington towpath passes under the arch. There was a time when this was one of the busiest places on the whole canal, home to a huge fleet of narrow boats. In the 1930s the Grand Union Canal Carrying Co. had their depot here, with fuel and supplies for their huge fleet of narrow boats. They carried goods for an array of local factories, listed in their promotional brochure as dealing with 'cocoa, chocolate, timber, jam, grain, potatoes, oils, gramophone record materials, coke breeze and engineering materials.'

The canal now begins a steady climb towards the Chiltern hills, with locks spattered along the way at regular intervals. Everywhere there is evidence of change, even on the old canal itself. Bridge 191 at West Drayton, for example, has obviously given trouble in the past and is now held firm by iron tie beams, still carrying the Grand Junction Canal name, while dates from 1912 to 1914 are pressed into the actual bricks in the bridge arch and towpath. It is hard to read the history of an area in surviving structures. No one could really guess that Cowley Peachy Junction, where the later Slough Arm leaves the main line, was once the terminus of the Paddington packet boat service, taking goods and passengers to London and back. The little aqueduct over the Fray's River has an old survivor alongside, one of the few remaining mile posts, now half buried in tarmac, and announcing that there are eighty-three miles to go to Braunston. Perhaps the most interesting survivor in Uxbridge is the wharf with wharf cottages and a boatyard, with a covered dry dock, the roof held up by cast iron pillars. It was here that one of the most famous of all carrying companies, Fellows, Morton & Clayton had their boats built and repaired. But these remnants of the past only serve to highlight just how much the world has changed since the canal was new. Suburbia has spread out from London along with those more recent phenomena, the industrial estates, which line the route, mostly sulking behind high fences. The 'lakes' that provide a watery gleam along much of the canal are also newcomers in the landscape, being mainly flooded gravel pits.

Change is inevitably most obvious in towns: Uxbridge has acquired a whole new character. The little market town is scarcely a memory as the office buildings multiply, some at least with a sense of style. A curvaceous block in green and cream has memories of art deco and something of the air of a stranded Cunard liner. The canal itself, however, remains rather aloof from change, and not necessarily the worse for that. Sprucing up has helped reinforce the architectural nature of bridges, by the use of whitewash for the main structure, but with the string course picked out in black, emphasising the contours. The little river which is a constant companion, more or less defines the line the canal has to take, but keeps getting in the way, so that the canal either has to hop across on a tiny aqueduct or briefly merge in a watery junction. Near Black Jack's Mill, now a family house, the river parts from the canal for the last time, tumbling away over an unusual U-shaped weir. Old industrial buildings mix with the new and one unusual feature appears above Springwell lock; a stone pillar with a weather eroded inscription. This is a Coal Duty stone marking the point at which duty was levied on coal shipments to London.

Rickmansworth provides the first distinctive canal settlement, largely because the waterway was built well to the south of the town centre. What makes the area particularly attractive is that this is a meeting point for three rivers, the Chess, the Gade and the Colne, through which the canal threads its way before setting out on a course up the Gade valley. But the Chess too was important, for it was navigable and joined to the canal by its own lock. So this was a natural spot on which canal development would occur. The very successful narrow boat builders, Walkers, had their base at Frogmore wharf; cottages and warehouses line the canal, and the local resident canal engineer had the best of it with his own very attractive house on

Black Jack's Mill epitomises the scenery of much of the Grand Junction, with its functional, yet attractive bridges scarcely disturbing the line of the horizon.

an island site between the Chess and the canal. Amazingly, London has still not quite been left behind, as one of the far flung arms of the Underground system reaches here, with a station and London Transport cottages.

Canal engineers often found themselves facing difficulties when they reached the land owned by some grand gentleman, and here there was not one but two to contend with, and both of the grandest – the Earls of Essex and Clarendon. They were not to be fobbed off with the commonplace structures found on the rest of the waterway, but wanted something more fitting to their estates. So Cassiobury Park, the Essex property, has an unusual pointed arch bridge and the Gothic theme is carried over to the otherwise plain lock cottage, which has been embellished with castellations. The Clarendons acquired an even grander bridge in their Grove Park grounds. A wide arch with a prominent keystone is topped by a cornice and balustraded parapet, complete with coats of arms: Even the towpath has its own arch through the abutments, flanked with pillars. At the canalside cottage, however, the only extra touch of grace comes with the round-headed windows. It may all have been done to appease aristocratic whims, but the results remain a delight.

A rare example of an ornate canal bridge. The canal passed through the estates of the Earl of Clarendon, and he demanded something grander than the common brick arch.

London is finally left behind as the canal passes through the new outer limit, the M25 orbital road, and then one is back with historic associations. Ovaltine had a factory here long before the present building was constructed, and they also had their own fleet of narrow boats with the company's familiar logo on the front. Approaching Hemel Hempstead there is another interesting example of a canal company having to make changes to mollify local interests. The original line between Apsley and King's Langley lay further to the east, rising through four locks, and taking water from the River Gade. This was not to the liking of the mill owners, who found themselves often running short of water. They forced the Grand Junction Co. to build the present line in 1819, using five locks instead of four to reduce the water demand, and briefly entering the river near the present railway bridge. There are few reminders of the change, apart from the numbering, with the addition of lock 69A.

Now the ridge of the Chilterns forms an inescapable obstacle up ahead. Jessop was no lover of tunnels – a view which was to be reinforced during the construction of this canal. So he opted for a solution that was to mark off the new generation of canal builders from their predecessors. He decided to drive through the ridge at Tring in a deep cutting. This might seem an easier option: digging a deep trench must surely be simpler than burrowing away deep underground, with all the dangers of roof falls and collapse. The difference, however, lies in the sheer volume of material to be shifted. The cutting was to be a mile and a half long, and at its deepest would be thirty feet below the natural ground level. The tunnel would have been at its broadest at water level, but the cutting slopes outward all the way to the top. At a rough estimate, five times as much material would have had to be excavated in the cutting as there would have been in a tunnel. When work started there were ten thousand men at work digging the canal, and a large proportion of them were concentrated here. Everything was done by hand, with pickaxe and shovel and plain black powder for blasting. A major problem was getting rid of the spoil so that work could go forward as quickly as possible. So the barrow runs were introduced. Planks were laid up the side of the cutting, and the barrows were hauled up by horses at the top, with the men walking the greasy planks and balancing the barrows in front of them. At the top, the barrows were tipped and the men ran back down, the barrows now behind them. It was a heart-stopping business that required courage as well as skill, and sometimes not even the toughest navvies proved up to it. There are no contemporary illustrations of the work on the canal at Tring, but the same system was used by Robert Stephenson on the adjoining cutting for the London & Birmingham Railway, and artist J.C. Bourne was there to record the scene.

This marks the summit, nearly 400ft above the Thames and water supply was clearly of immense importance. The original plan called for a feeder that would tap the numerous springs around Wendover, but it was decided that if a channel had to be cut anyway it might just as well be made big enough to take boats to the town. So, the Wendover Arm was duly built and opened in 1797. It was not a huge success, largely because of seemingly insoluble problems of leakage through the chalk, which was sometimes so bad that there were times when the feeder that was supposed to be supplying water to the main line was actually draining water out of it. So, between 1806 and 1817, a group of three reservoirs was added – Marsworth, Tringford and Startop's. Two of these are below the level of the main line, so water had to be pumped up, originally by a steam engine, housed in the specially built Tringford Pumping Station beside the Wendover Arm. More reservoirs, Startopsend and Weston Turville, were added not just to supply the canal, but to make good supplies to mill owners. To add to this busy scene, the Act of 1794 that authorised construction of the Wendover Arm also gave permission for another branch which would run into the heart of Aylesbury. It was originally thought of as part of a much grander scheme that would eventually link through to the Thames at Abingdon, but Aylesbury was to prove the end of the line. It was still a major undertaking with sixteen locks, but these were built narrow to save both water and money. It ends at a basin with just one surviving warehouse. It was a success however in helping to attract industry to the old market town, when a condensed milk factory was built close by the canal in 1870. With a long summit level in a deep cutting, two branches leaving the main line within a distance of less than a mile and reservoirs and pumping stations, this was just about as busy a spot as you could find on any canal. It was an obvious place for a maintenance yard, and it was duly built at Bulbourne. A hamlet grew up alongside the yard, but it is the workshop complex that demands attention. Bulbourne yard has altered surprisingly little over the years, and among the essential jobs, the making of wooden lock gates continues in a tradition that stretches back to the earliest years of the canal. The buildings are nineteenth century, functional and plain but with what was then a modern touch in the cast iron windows, and with just a hint of decoration in the lighter brick that emphasises the arches over windows and doors. There is a pleasing note of frippery in the ornate little clock tower.

The canal at Marsworth winding round the reservoir, built to supply water to the Tring summit level.

After that flurry of excitement things become altogether easier as the canal heads off on a moderately straight course for the upper end of the Ouzel valley, with Leighton Buzzard looking across the water to neighbouring Linslade. This seems a peaceful rural scene, but the arrival of the canal spurred on a new industry; sand extraction for the building trade. The pits have now either been filled in or flooded, but at its peak sand carriage on the canal reached a very respectable 40,000 tons a year. The artificial waterway mimics the convolutions of the River Ouzel as it winds down a narrow valley. A level is kept as long as possible, but eventually there is a plunge down through the three locks at Soulbury. This forms a break with the established pattern of single locks spread evenly along the line which has marked the progress from Bull's Bridge, and to ease the traffic on its way a second set of narrow locks was built alongside the wide for the use of single narrow boats – not that the boating families necessarily minded a delay here as one of many canalside pubs was built right next to the locks.

The interruption over, the canal continues through what would have been open country, very open in this case for the region suffered major depopulation in the Middle Ages, leaving deserted villages that are now no more than humps and bumps in the fields, while great houses have vanished, remembered only by the faint outline of an old defensive moat. There are still these reminders close by the canal, but the landscape was to be filled in again in more

The three locks at Soulbury are conveniently grouped together, and provide a natural break in the journey – just the place for a canalside pub.

Bulbourne maintenance yard.

The forge at Bulbourne, little altered since the building was completed.

recent times. There is a little village called Milton Keynes just to the east of the Ouzel, but its name has been borrowed and given to the new city. So the canal now skirts an extensive area of urban development, its countrified air vanished. Because of its new surroundings, it is no longer clear that the canal is also going through a change. The Ouzel is abandoned and the progress north temporarily halted as the Grand Junction turns west towards the valley of the Great Ouse. Unfortunately for the engineers, the Great Ouse heads from west to east, blocking the way forward. So the canal is forced to come to terms with changed circumstances. It turns east at Linford Wharf, starting point of the long abandoned Newport Pagnell Arm, closed in 1864. A railway was built over the line of the old canal, but that too has closed and is now used as a footpath.

The canal keeps to the side of the river valley, waiting until a bend in the river provides the appropriate crossing point. The later railway engineers who came this way took a more direct line, and because they had reached a point more or less halfway between London and Birmingham, they decided to build their locomotive and carriage works here at Wolverton, and then went on to build a whole new town for their workforce, New Bradwell. Rather surprisingly perhaps, very little of this impinges to any great extent on the canal, where the most obvious 'industrial' reminder comes in the shape of a restored windmill by Bridge 72.

The crossing of the Great Ouse could not be postponed for ever, but there was a great deal of argument over how it was to be achieved. The river valley is broad with gently sloping sides. The first option to be considered was for four locks to bring the canal down to the river, which could then be crossed on the level, and four more locks to bring it up again on the far side. There were two objections to this. Firstly, it would be very wasteful of water, and some form of back pumping would have been needed. Secondly, the river was very prone to flooding. Barnes proposed an alternative solution. Embankments would bring the canal to the riverside, with the final leap across on a conventional masonry aqueduct. The trouble was that by the time this proposal was put forward, work was continuing on the canal from both the north and the south, and such huge earthworks would have delayed the opening of the through route. Jessop proposed a compromise, to build temporary wooden locks which could be used until the more significant works were ready. There was another factor in the equation. To the north, where tunnelling was in progress at Blisworth there were more delays in completing the connection by water. But a temporary solution had been found by building a tramway across the hills towards the Great Ouse. By adopting Jessop's suggestion a through route could be opened up all the way from Braunston to London, even if the horses would be pulling trucks on rails instead of boats on a canal for a part of the journey. The idea was accepted and the route duly opened. All seemed to be going well on banks and aqueduct. There were grumbles about the quality of workmanship, which was not unusual, but in any case money was scarce and everyone was keen to get on as fast as possible and start collecting revenue. Then reports came in that all was not well. First the embankments began to crumble and slide and had to be repaired. More ominously, the arches of the aqueduct were misaligned and cracks began to appear, but nothing was done until the inevitable disaster occurred. The entire structure collapsed, blocking the river and raising fears of widespread flooding. At last a sensible decision was reached. The banks were made good and a cast iron aqueduct designed. It is not a graceful structure. The ends of the banks were revetted to create solid abutments and a central pillar erected in the river. The cast iron trough itself is constructed of angled cast iron plates, strengthened by curved iron side plates forming arches. The problem was solved and the aqueduct still stands.

This bridge at Braunston clearly shows the standard design for Grand Junction bridges, with the lines emphasised by a prominent string course and parapet.

Once across the aqueduct the canal arrives at another old junction at Cosgrove. What is now the Cosgrove Marina was once the start of the arm to Buckingham, with a branch to Old Stratford. They were effectively closed to traffic in 1900, even though they were not officially abandoned until 1964. The canal engineers could now enjoy the luxury of a lock-free section as they reached yet another river valley, the Tove, a tributary of the Great Ouse. It is only a short-lived respite, for the ground begins to rise quite steeply. Seven locks take the canal up to what is perhaps the best known of all canal settlements, Stoke Bruerne. A canal museum was established here in 1963 in a former corn mill, a simple building of three storeys with loading bays over the wharf. But it is not just this that makes Stoke Bruerne so attractive: in fact, the museum is here precisely because the whole area is so appealing. Former lock-keeper Jack James was famous for the way in which everything around the locks was kept so immaculate and for the collection of canal memorabilia he kept in and around his cottage – and which was to form the core of the museum displays. This is a spot where often disparate elements come together in a most satisfactory way. Mill and houses are mostly built of the local stone, which is lias tinged with iron, that gives it a mellow, golden-brown colour. A brick house might seem dull in such company, but here the single hipped roofed house is enlivened by the use of diaper work, a chequered pattern created by using different coloured

A rare sight on the waterways: the steam narrow boat President *with the butty* Kildare *entering Stoke Bruerne top lock with the waterways museum ahead of them.*

The Anatomy of Canals

The toll office, Braunston: the Horseley Iron Works bridge in the foreground shows its construction, as a series of iron plates bolted together.

bricks. Roofs are mainly of slate, not a local material by any means, but one that became widely available largely as a result of the improved transport offered by the canal itself. The exception is the old pub across the water, which is thatched. Like the other locks in this section, the locks had been doubled, but the extras have mostly been filled in or abandoned. Here, however, the second lock has been turned into a kind of open-air museum exhibit, containing a boat weighing machine that started life on the Glamorgan Canal. The final feature that completes this appealing scene is the double-arched bridge across the tail of the top locks. Those whose main interest is in the history of the place can discover the line of the old tramway which briefly formed a vital link in the system. It ran on the far side of the canal from the present towpath. This was always an important place, especially in the days of horse-drawn boats, for it was here that boats and horses would part company – the horses to be led over Blisworth Hill, the boats to be taken through the 3,057 yard long Blisworth Tunnel. While the advantages of a wide canal are obvious, the disadvantages appear here. The tunnel has never had a towpath, so boats had to be legged through. In order for the men to reach the sides with their feet, special boards had to be pushed out at each side of the boat, the boatman lying on his back on one and, where necessary, a professional legger taking the other. The system remained in use until the advent of powered tugs.

There had been much argument over whether to build a tunnel or an open cutting as at Tring. Here it would have been dug across the brow of the hill, reached by locks at either end and kept supplied with water by steam pumps. In the end, the tunnel was chosen. It was to be 16ft 6in wide, with an irregular cross section, the crown 11ft above water level having a greater radius than the invert 7ft below the water. To make the work easier, drainage headings were cut before the main work of tunnelling began, after which the men worked out from the bottom of nineteen vertical shafts as well as from the two ends. The work took three years and was completed in 1805. Originally, the whole tunnel was brick lined, but continuing problems called for a major rebuild and boats now pass through what is in effect a giant circular tube, constructed of linked circular concrete sections. This work was completed in 1984. The only reminders of the old workmanship are the circular brick towers lined up on top of the hill, marking the working shafts that were retained for ventilation.

From the far end of the tunnel, a cutting now leads to Blisworth itself which, if it lacks the more obvious charms of Stoke Bruerne, is certainly not without interest. Here, however, the mill of 1879 has not enjoyed the careful restoration that has made the Stoke Bruerne mill such an attraction. There is also a reminder of the years during the tunnel construction, when the gap was filled by the tramway. Pickford's, one of the major carriers on the canal, built a transit depot where goods could be stored for transfer between boats and trucks, a simple building with a canopy over the wharf. There is also a curiosity in the name given to Bridge 50 –

The former army barracks and stores at Weedon. This was a Royal Ordnance depot, built during the Napoleonic Wars, and was to have been supplied by canal.

Looking not unlike a windmill tower, this is actually a ventilation shaft, rising above Blisworth tunnel.

Candle Bridge. It recalls the time when boat crews stopped here to buy candles to light their way through the tunnel. After Blisworth, there is just a short way to go before the next important point, Gayton Junction. This is the start of the Northampton Arm, one of the branches planned from the start, and like the others, built with narrow locks to save both money and water. Unlike the other branches, however, this is no dead end, but connects with the River Nene, providing access to the complex system of waterways that spreads across East Anglia. It is only five miles long, but crams in seventeen locks, thirteen of them grouped together in the Rothersthorpe flight. There is evidence of cost cutting in the use of wooden lift bridges, instead of permanent brick bridges. There is a good example at the tail of Lock 5.

Once again the canal is able to keep a level by contour cutting round the edge of the very undulating landscape to the south, while heading towards the upper reaches of the Nene. The two waterways are only a short distance apart by the time they reach Weedon, a point which authorities noted was very near the centre of England. The waterway arrived here, from the north, in 1796 when the threat of invasion by Napoleon was at its height. It seemed to be the ideal place for a major ordnance depot, as far as it could be from the coast and any invasion point, and with excellent communications through the newest and best of the country's waterways. So the Royal Ordnance Depot was constructed at the end of a short arm which ended in a basin at the heart of the complex, and which could be closed off by a portcullis. Barracks and stores were laid out with true military symmetry and there was a Pavilion where, according to popular but unconfirmed accounts, the King and the court would be settled if the French landed and threatened London. The Army finally left in 1965, untroubled by invaders.

It is difficult to see this elegant pair of bridges at Braunston turn as the product of a new mass production technique, but that is just what they represent.

Once past Weedon, three transport routes converge, with the canal squashed in between the railway and the M1. Where the railway engineers were able to take a direct line through major earthworks, alternating high embankments and deep cuttings, and the motorway engineers simply went up and down hills as they chose, the canal engineers had to reach a higher level, through the Wilton and Buckby locks. At the top of the second flight, an isolated house, once again built of the local rich, warm ironstone, marks the arrival of another junction. Here the Leicester Line leaves the Grand Junction, not an arm but an independent waterway built by a quite separate company, which will be dealt with in the next chapter. And now the Grand Junction faces its last major challenge. It enters a summit level, fed by water from a reservoir at Daventry, and disappears into Braunston Tunnel, built before Blisworth, and not as long at 2,042 yards. But it still represented a formidable challenge, partly because quicksands were encountered which had not registered in the trial borings that had been made before construction began. Control of the works was less than perfect, with two sets of contractors working away, but sadly not on the same headings, as a result of which the tunnel has an interestingly serpentine bore.

From the end of the tunnel, the canal drops down through six locks to the once busy canal settlement of Braunston. The most striking feature is the tall chimney, still bearing the letters GJC and the date 1897, which stands above the hipped roof of the pumping station, a typical building of its period, making decorative use of contrasting red and blue bricks. The end of the line is marked by the Stop House, the toll house by the junction with the Oxford Canal. It once stood beside a stop lock, that marked the boundary between the two canals, but this has now been removed. Here a feature last met right at the beginning of the canal reappears, a cast iron bridge by the Horseley Iron Works, a standard casting that is duplicated at Braunston turn. The advantage of having a standard set of castings that could be delivered like an Ikea flat pack and erected anywhere are obvious, but happily this is also a very pleasing design. The arch curves gently and elegantly across the water, topped by a parapet with an open pattern of hexagons. It is a perfect marriage of form and function.

This was as far as the Grand Junction was ever intended to go. The plan was for the journey to Birmingham to be continued along the northern section of the Oxford Canal. But long before work was completed, new canals were being planned and built which would give new options, open up new routes and lead to a major expansion in the whole Midland canal system.

3. Towards a Grand Union

At the same time as the Grand Junction Bill was making its way through Parliament, another group was promoting a canal from Birmingham to Warwick. It now required no great leap of the imagination to see that this offered the opportunity to make yet more improvements in the London-Birmingham route, by means of another new canal from Warwick to the Grand Junction. So, in the following year, 1794, another Act was passed, to link Warwick to Braunston. The Oxford Canal Co. might have been expected to raise objections, but they were in the happy position of being able to effectively block the plans if they chose, and were quite happy to allow through traffic – at a price. In the event, the junction was moved to Napton, which meant that the Oxford Canal would take traffic along the new system between there and Braunston. It all made a great deal of sense, but the new canals were not planned to the same generous gauge as the Grand Junction. Having their origins in the Birmingham system they were designed to fit in with the narrow locks that up to then had become standard throughout the extensive Midlands network. That, however, is not the end of the story as far as these canals are concerned. In 1929, they and others joined with the Grand Junction to form the Grand Union Canal Co.

Work began at once on unifying the system, which involved replacing the narrow locks with broad locks, dredging the canal and piling or walling the banks to increase the cross-section. The new locks appear very different from their predecessors. Not that the fundamentals have changed, but the paddle gear is enclosed, protecting the moving parts from the elements, and what appear to be rows of inclined metal posts march up the long lock flights, such as that at Hatton. The other change is less immediately obvious, but the engineers turned away from traditional brick or stone for the lock chambers and went for the modern alternative of concrete. The old narrow lock chambers were mostly kept and adapted for use as overspill weirs.

The original idea had been to work the improved canals with broad beamed barges, but that would have involved even more expense in widening bridge openings. It was soon realised that any gain in cargo carrying capacity obtained by using bigger boats would be cancelled out by the extra delays as the craft tried to pass each other. The logical decision was to continue what was already a common practice of working narrow boats in pairs that could fit side by side in the locks, and with one towing the other, they presented no problems at bridges or in narrow sections. The result of all this was that many structures were rebuilt, not just the locks, and although the northern end of the Grand Union retains the line laid down in the eighteenth century, the details and many of the buildings only date back to the early twentieth century.

The original Act for the Warwick & Birmingham Canal specified rather vaguely that it would 'terminate at or near to a certain navigable canal in or near to the town of Birmingham, called the Digbeth Branch of the Birmingham and Fazeley Canal Navigations'. The wording suggests that not all differences had been resolved between the two companies who saw the benefits of co-operation but were unwilling to agree to anything that might harm their own special interests. As usual in these matters, the stumbling block was water supply, which was finally resolved by the insertion of a stop lock, the Warwick Bar, which the Birmingham & Fazeley were empowered to close at any time when the water in their canal dropped below a certain level. The proprietors of the Warwick & Napton Canal had no such problems with their new neighbours, and the Act made it clear that these were equal partners, the two canals meeting on the level and sharing water. Rather than dealing with these canals in chronological order, this account will start at Braunston, picking up the account from where we finished in the last chapter.

The Grand Junction came to an end at Braunston, and from here the Oxford continued the route south to Napton, a distance of five miles. This was a very expensive five miles, for the Oxford demanded high tolls for its use. Inevitably water supply was soon on the agenda. The Oxford, not unreasonably, refused to take any responsibility for the newcomers' problems and had no intention of seeing their own precious water being drained off to feed someone else's canal. In the event, the Warwick & Napton had no choice and work began on new reservoirs at Napton Junction.

One of the curiosities of this canal is the bridge numbering which starts at 17. In fact, this simply reflects the original plans which called for a junction at Braunston, where presumably

It now seems a romantic idyll, but this was once a commonplace sight on Britain's waterways: this horsedrawn narrow boat is near Ivinghoe.

This bridge at Bascote is an excellent example of a substantial road bridge, which comes complete with a long access ramp to the towpath.

bridges 1 to 16 would have appeared along the way. Once past the reservoirs, the descent towards Warwick and the Avon begins with the three Calcutt locks. Those who come this way, and who remember their geography lessons, will recall that Birmingham stands on top of a plateau, and the more the canal sinks down at this end, the more it will have to rise at the other. There is also a hint of what lies ahead, with Bridge 20 named Gibraltar Bridge after nearby Gibraltar House Farm, suggesting that rock might well be an important feature in the landscape. And sure enough an arm duly appears, leading off towards a wooded hill, scarred with quarries. Then the canal reaches the first long flight of locks at Stockton. They are set close together for ease of working, and they come with a 'new' lock cottage, one that only dates back to the 1930s and certainly looks the part: it would not be out of place in Acacia Avenue, Suburbia.

It is at Stockton that the stone themes come together. First a canalside pub gives the geology lesson, the Blue Lias. This is the name of a form of limestone, not really blue, but a kind of slate grey. It is not particularly useful for building, as it is usually found in small, irregular pieces and does not weather well. It is, however, just the thing for making cement. A short derelict arm leads off on one side of the canal towards old quarries, while a more distinct arm leads south to what was once Kay's Cement Works and is now Rugby Cement, easily identified by the tall chimney which is a prominent landmark. For a time this was an important part of the canal trade, but just beyond Long Itchington, the railway branch line that took over much of the cement trade strides boldly across both the canal and the River Itchen. It rather detracts from the efforts made by the canal engineers, who brought the canal up to the Itchen on a high embankment with just a modest aqueduct for the actual river crossing. Changes did not end with the 1930s improvements, and a new generation of

The skew bridge at Budbrooke, built across the canal at an angle. This involves some complex brick laying under the arch.

engineers has kept faith with old forms. The road bridge Number 27 has a stepped ramp, corbelled corner and a decorative effect is created by the use of bands of red and blue bricks.

The line of the canal is somewhat wriggly, winding through the undulations. There are no very high hills to bother an engineer, but a rise of just twenty or thirty feet represents a considerable obstacle for a waterway. If the diversions are not too big, there is a lot to be said for going round rather than through or over in expensive earthworks. But, at some stage locks are inevitable as the land slowly dips towards the main river valley. Bascote locks provide an interesting variation. There are four of them, but in the rebuild the first two were run together, with the bottom gates of the top lock also being the upper gate of the next. The other locks are separated by widened pounds which act as mini-reservoirs. The scene is completed by a toll house and lobby at the top of the flight. Now there is a steady scatter of locks leading down to the Leam valley and Royal Leamington Spa. There is nothing very royal or even very spa-like in the town that borders the canal: not much hint of polite gentility here.

A map of 1783 shows a small rural community, a smattering of cottages to the south of the river crossing. It was only over the next decades that a new centre of development appeared on the opposite side of the river. There were now two quite distinct Leamingtons, one of grand houses, broad streets and fashionable squares: the other of narrow streets, mean terraces and industry. This was the Leamington that developed along the axis of the canal, and the industrial world soon appears. At bridge 35 the pyramidal roofs of a canalside building suggest a brewery, and this is indeed the former Thornley Brewery that turned out its last pint in 1968. Bridge 38 has a site of great importance, the wharf of the Eagle Foundry. There are hints of other works, and a map of 1814 makes clear just how busy Leamington was when the canal

One of the most daunting sights on Britain's canals, the long flight of locks heading up the hillside at Hatton.

was new and what an important role it had to play in this aspect of the town's development. It also shows how the fashionable spa pulled up its skirts and kept its distance.

A similar story can be told as the canal slides off towards Warwick. The transition is marked by two aqueducts. The first, a later addition, takes the canal across the Oxford & Birmingham Railway and was built in the 1850s, with an iron trough and iron balustrading carried on brick arches. Almost at once this is followed by a crossing of the Avon on a conventional sandstone aqueduct, with three arches and with two of the piers set in the river. The canal continues on an embankment with views out over the suburbs and with ample evidence that Warwick, too, had a busy and varied industrial presence when the canal age was at its height. Among the buildings that can be seen from the high viewpoint of the bank is Emscote Mills, built in the eighteenth century with a typically regular array of segmented arched windows. It has the appearance of a textile mill, but in later years was used for manufacturing gelatine. Everywhere there are signs of a once busy commercial traffic at wharves and boatyards. Now, having crossed the Avon it is time to begin the long climb up to Birmingham. It starts modestly with the two Cape locks by the Cape of Good Hope pub – but whether the locks took their name from the pub or vice versa is uncertain. The end of this canal is now very close but easily overlooked. A short arm leaves the main line by Bridge 51. This is, in fact, the final section of the original Warwick & Birmingham, which then joined the Warwick & Napton at what became Budbrooke Junction.

The canal began at Saltisford, but the old basin has been filled in. The arm itself was abandoned, but restored in the 1980s. It is not easy to see now, but the basin was once an important focus for industrial development. Not surprisingly, on returning to the main line, there are no immediately obvious differences to show that one has left one independently owned canal and joined another. This is partly because the two were planned from the first in a spirit of co-operation, and also because any differences there may have been were smoothed away in the creation of the Grand Union. The engineers on this part, however, faced far greater problems than those of the Warwick & Napton. There is no gentle climb here, but a charge up the hill through the Hatton Locks, twenty-one of them lifting the canal 146ft in just a couple of miles. This is not the longest lock flight in Britain, but it is certainly one of the most impressive. The first part lies on a curving line, but then the canal straightens up revealing a daunting array of black and white balance beams marching away into the distance. The engineer, William Felkin, is not one of the better known canal builders, but he showed throughout that he had a thoroughly modern outlook and a good understanding of how to organise things to ensure efficient operation. Locks are kept close together for ease of operation, a lock cottage sits halfway up the flight so that there was someone on hand to control matters and if things did go wrong there was a maintenance yard available at the top, with a toll office. The yard itself has a sombre appearance, built of dark blue brick, and it might

The lock office at Hatton Bottom Lock. The tiny building is no more than a simple extension to a standard lock cottage.

The canal rarely impinges on the elegance that made simple Leamington into Royal Leamington Spa, but this is one of those places.

seem odd to have a toll office half way up a canal, but there is a junction just a short way ahead. Water conservation on the flight is helped by the use of side ponds. The wide Grand Union locks use prodigious amounts of water: it takes over 50,000 gallons to fill the average broad lock. When in use, the side ponds act as small reservoirs. When emptying a lock, the water runs into the side pond until levels are equal, then runs away downhill as usual. The water in the side pond is now available to fill the lower part of the lock for the next user.

There are now some nine miles of lock-free canal up ahead, but that does not mean that the engineers had an easy time of it. The Ordnance Survey map for the area shows a higgledy-piggledy mass of contour lines, like a child's scribbles, that translate on the ground into a landscape of hillocks and hollows. It must have been extraordinarily difficult to decide on the best line to take: where to wriggle round obstacles, where to go through or over them. Felkin did a bit of each. From Hatton top lock the canal swings round to confront the hill of Shrewley Common. It heads down into a deep cutting that ends in the 433 yard long Shrewley Tunnel. It has no towpath, and in pre-motor days boatmen pulled their boats through by means of handrails set into the tunnel sides. Horses were led over the top, but completed their journey through a short horse tunnel of their own, emerging next to the canal tunnel mouth but at a higher level. A long ramp leads back to the bottom of another deep cutting, this time with exposed rock, showing variations between marl and sandstone.

Having pierced one hill, the engineer was now faced by another at Rowington. He contemplated a second tunnel, but opted instead for a technique that was typical of the period – cut and fill. There was a valley to cross before reaching Rowington hill from the south. As the cutting was hacked out, through the summit, so the spoil was removed and used to build up an embankment in the valley. It now seems a very obvious approach, but it was only in the 1790s that the technique became widely used, and it was very popular with a later generation of engineers who built the railways. You can see it on the line that runs by the canal, where the method is used over and over again. Once through the cutting, however, further earthworks were avoided by taking a contour round the hillside.

The next point of interest is Kingswood Junction, where a short arm leads to the Stratford Canal, and most of the interest lies on the latter as described in Chapter 1. Now the canal turns north for a fairly easy passage to the five Knowle locks. These replace the original six narrow locks and were the last to be altered by the Grand Union: by the time the next locks are reached we shall be back with the narrow locks of the Birmingham system. That is still a long way off, and for the next few miles the pattern of cut and fill, alternating with the occasional wide swerve round an obstacle, continues. There are a few special points of interest. The crossing of the valley of the Blyth is accomplished by a combination of embankment and short aqueduct over the river itself, which also acts as a drain for the canal. An overspill weir carries water from the canal cascading down through the trees to the Blyth. A sharp bend follows with a stout iron post at the apex. Tow ropes were passed round this to help negotiate the bend, and over the years they have dug deep grooves into the metal.

The canal was once rural for most of its length, but now suburbia has spread out from Birmingham, and the built up area begins around Bridge 78. It soon disappears from view as the canal heads off through a deep cutting, but then it emerges on the inevitable accompanying bank for a rooftop panorama. Nothing is as it was – new houses, new factories and even familiar canal features have been given an update: mile posts have been replaced by kilometre posts. The next river crossing is less than dramatic, not so much an aqueduct as a culvert through the bank to carry the River Cole. Then comes Camp Hill, still with an air of the working past with its basin, waterways depot and warehousing. Then there is a brief interruption to the climb up to Birmingham as the Camp Hill locks, still the narrow locks as originally built, lead down to Bordesley Junction. This is a comparative newcomer, where the nineteenth century Birmingham & Warwick Junction Canal was built as a shortcut to the Birmingham & Fazeley. There is a two-arched aqueduct over the culverted River Rea which leads to Warwick Bar and the turn off to the Digbeth Branch. It is an interesting area of old warehouses, one of which was used by Fellows, Morton & Clayton. It is indicative of changing attitudes that Birmingham, which neglected its canals for so long and acquiesced in the destruction of so much of the old, should now have made this a conservation area. It deserves it.

This marks the end of this section, but not the end of the Grand Union. There was another Grand Union before the 1929 amalgamation. This one was approved by Parliament in 1810, though it was not for a brand new canal: it was an attempt to resolve an impossible situation that had been dragging on for years. The story begins at the height of the mania years in 1793 when an Act was passed for the Leicestershire & Northampton Union Canal. William Jessop had already been at work in the area in creating the Soar Navigation through Leicester. The new canal was, as its name suggests, intended to link the Soar to the Nene at Northampton and also to join the Grand Junction through the Northampton Arm. Jessop was again called in for his professional opinion, but was too busy to oversee the actual construction. This was to be a broad canal, with 14ft wide locks, and at first things went reasonably well. By 1797 they had advanced some seventeen miles from Leicester and had reached Debdale near Gumley. Then the money ran out and everything came to a halt. Various engineers were consulted. James Barnes arrived from the Grand Junction and advised that the Northampton route should be abandoned and a new line taken down to Braunston. No one seemed interested. In 1802 he came back with another proposal, still for a direct route to the Grand Junction, but this time taking a shorter line to Norton. The company, showing their usual decisiveness, agreed to nothing except that they needed another opinion. This time it was Thomas Telford's turn, and he produced yet another variation – Norton would still be the objective, but the canal would go there via Market Harborough. The company finally agreed that this made sense and work restarted, but by 1809 they had still only staggered on as far as Market Harborough where they stuck once more. By now, the Grand Junction had lost all patience and went on to promote the Grand Union Act to get the job finished. This allowed for a line through Foxton to Norton, but to save money it would be built as a narrow canal. At last, in 1814, after twenty years of a little work and a great deal more inaction, the line was complete. Leicester and the other waterways that served it were joined to the Grand Junction.

Strictly speaking, what is now known as the Leicester Arm of the Grand Union falls outside the scope of this volume, because of its late construction date, but for the sake of completing the story it will be dealt with here. It is an extraordinary canal in many ways. One tends to think of the main line of the Grand Union as epitomising 'modern' canal engineering practice: taking a direct line, even where this involved extensive earthworks, tunnels and aqueducts. Look at the Leicester Arm on a canal map, however, and one seems to be right back in the age of Brindley. Once again, one has to turn to a map showing contour lines to appreciate the problems faced by the engineers. Little brown worm threads wriggle all over the Ordnance Survey maps, without so much as a hint of a useful river valley to provide a respite. In fact, this canal offers a daring solution to a difficult problem. From Norton Junction, the canal climbs steeply up to a level, which is then maintained all the way to Foxton, a distance of twenty-two miles. Then there is an even steeper descent, for the final

Stockton Locks show the results of the great improvement scheme begun by the newly formed Grand Union Canal Company in the 1930s, replacing old narrow locks.

Wide locks are left behind when turning off the main line onto the Northampton Arm. This is the thirteen-lock Rothersthorpe flight.

run to Market Harborough. In these circumstances, the wavering line seems amply justified and, as we shall see, it involved a good deal more than simple contour cutting.

The start at Norton Junction is innocuous as it heads for the break in the hills known as the Watford Gap. Generations of transport routes have headed for this convenient point, from Roman Watling Street, through the canal age to the railway age and on to the modern M1. How many motorway users who stop off at the M1 service station here are aware that right alongside is one of the area's most interesting canal features? This is the climb to the summit, through seven locks, the first five of which are arranged as a staircase. Everything is geared to efficiency, with no frills: even the footbridges at each lock are of the plainest design. Water conservation is tackled by the use of side ponds. Having reached the desired level, an obstacle at once appears, the hill at Crick. Too big to go round, too high for a cutting, the only possible answer that could be found was a tunnel. It was to prove extremely difficult to construct as a result of the nature of the ground, an unstable mixture of clay and sand, always liable to slippage. When Robert Stephenson's men came to build the nearby Kilsby Tunnel for the London & Birmingham Railway, they nicknamed this stretch of land 'Quicksand Hill'. The canal workers would not have argued with that description. This was no minor concern, for it was nearly a mile long and, although it has no towpath, was built wide enough to allow narrow boats to pass.

Now the wanderings and the weavings begin. In one way, this was not a problem, for there are few settlements along the way to provide cargo. Welford, the one place of any size, had its own arm. Elsewhere the surveyors could simply find the easiest route with no other considerations to distract them. One more long tunnel was needed, however, at Husbands Bosworth. Then at Foxton, the point was reached where it was time for the big descent, an

The very modest opening to the half-mile long Saddington tunnel on the Leicester Arm of the Grand Union.

The former Donisthorpe cotton mill with its pedimented frontage, displays its Georgian heritage. The factory bell hung in the small cupola.

even steeper drop than that at Watford, a total fall of 75ft. This called for two five-lock staircases, with a passing place in the middle. Even with this provision, Foxton proved a major bottleneck on the canal, which was in time to lead to an interesting solution, which will be dealt with in the final volume. Once past Foxton, the meanderings continued with the biggest deviation of all, the canal tracing a giant 'U' on the ground as it heads off to Market Harborough. It ends here in a large basin, that has recently been redeveloped.

The remainder of the route to Leicester brings a change in lock size, to allow 10ft wide barges to pass from the River Soar. The line here is altogether more conventional, still somewhat wriggly, but now with a steady spattering of evenly spaced locks. Eventually, the canal makes its way into the Soar valley and enters the river at Aylestone – an interesting junction, made very close to a large weir. The Soar Navigation continues the route through to the Trent, passing through the centre of Leicester, and providing a good opportunity to see some of the city's older industrial buildings, which include the handsome former Donisthorpe cotton factory with its neat cupola. It is only by seeing this concentration of industry that one can make any sense of a canal that otherwise has spent its time apparently wandering aimlessly all over the countryside.

Towards a Grand Union

In spite of adaptations to modern use, the former warehouses at Market Harborough retain their functional appearance.

The imposing Foxton staircase, with side ponds terraced into the hillside to keep the locks supplied with water.

The Leicester Arm wanders through a very rural landscape, with the Crick Marina in the distance.

The tail of the lock on the Welford Arm is crossed by a simple footbridge.

A simple hand crane provides the dominant theme at Kilby wharf on the Leicester line.

4. Waterways to Wales

Wales had been ignored in the first wave of canal building. This was to change dramatically in the 1790s, largely because of a tremendous surge of activity in the iron industry. It was the introduction of coke instead of charcoal as the fuel for the blast furnaces that provided the impetus, for in South Wales, and to a lesser extent in parts of North Wales, both iron ore and coal were available, together with limestone which was used as a flux to remove impurities in the form of slag. It made sense to build furnaces close to the sources of raw materials, after which all that was needed was an appropriate transport route to take away the finished product. Canals were the obvious solution in the eighteenth century. Developments in the north were however different from those in the south. In the north, canals were considered as an extension of the established and still growing English network. There was also another important use for limestone to be considered, turning it into lime as a fertiliser, a use which related to the rural economy rather than the new industrial world. In the south, geography resulted in a very different pattern, with new focal points in the south coast seaports of Cardiff and Swansea. Much of the new industry developed in the valleys that generally run in a north-south direction in more or less parallel lines. So the two sets of canals developed quite different characters, and this chapter will concentrate on the north.

What is now the best known of the Welsh canals had its origins in what had seemed to be a languishing failure, the Chester Canal. Begun in high optimism, all that had been built was a line from the Dee to a point near Nantwich, where it remained in doleful isolation. Two things were obvious. The Dee was a failing river, as far as trade was concerned, steadily losing traffic to the Mersey and the burgeoning port of Liverpool. The Chester Canal would never prosper unless it connected with the kind of industrial areas that would provide profitable cargoes: remaining in its present form, serving purely rural areas could only end in closure. An extension was needed, but in which direction? There was a great deal of debate over what line should be adopted, and plans were still being changed even after work had begun. Eventually there would be new connections at both ends of the Chester Canal. One, the simplest route, would head north to the Mersey; the other would go west through Shropshire to North Wales to tap into important sources of coal, limestone and iron and, it was hoped, connections would also be made through to the Severn. En route, this latter line was to pass through the little market town of Ellesmere, which was to give the canal a name, the Ellesmere Canal, and also name the new port that would develop where the canal joined the Mersey, Ellesmere Port. The old name for the canal is scarcely remembered these days and this reflects change in use and fortune. First, it was incorporated into the Shropshire Union system, which swallowed its identity. Then, industrial use faded away and pleasure boating took over, and the canal received a new name that would certainly have surprised its builders.

When the canal was under construction, the usual problem of water supply had to be addressed. Near the end of the run into North Wales, the canal was to pass high above the River Dee. The engineers traced a route back to a point on the river, above the level of the canal. There they constructed a weir, now known as the Horseshoe Falls, and water was diverted away down a watercourse carved into the steep hillside, for a distance of some six miles. As this feeder passed close to the town of Llangollen, it was decided to make it wide enough – though only just – to take boats. The feeder was a navigable canal, and we now think of the whole canal as having been designed with the express purpose of taking boats to enjoy the delights of Llangollen, so the Llangollen Canal it has become.

The chief engineer for the canal was William Jessop, who, by the time the Act was passed in 1793, was heavily committed to work all over the country. The day-to-day overseeing of construction was entrusted to the 'General Agent, Surveyor, Engineer, Architect and Overlooker of the Works', the Scotsman Thomas Telford. His career up to this point had been remarkable enough. A shepherd's son, he was apprenticed as a stonemason, moved to London to seek his fortune, found work on such prestigious projects as Somerset House, and attracted the attention of a number of influential men. As a result in 1786, at the age of thirty-one, he had risen far above the ranks of stonemasons to be appointed Surveyor for the County of Shropshire. This offered wide ranging experience from bridge building to acting as architect on important buildings. Two of his churches survive, at Bridgnorth and Madeley, and though they are competent exercises in the rather chilly classicism of the day, they lack real originality. He never lost his love of architecture, but he was to find fame as an engineer. The question that has never been completely resolved is who should be given the major credit for the Ellesmere, Jessop or Telford? Some points are clear. The overall planning of the route, the all important decisions over how to lay out the line of the canal – where to build locks, banks, cuttings, tunnels and aqueducts – these were all essentially down to Jessop. Some of the details which still delight us today, particularly houses, cottages and offices, bear the mark of Telford

The Ellesmere Canal begins its climb up from the Cheshire plane to the Welsh hills at Hurlestone Junction.

the architect. But when it comes to the most famous structure of all, the great aqueduct of Pontcysyllte, the only safe answer is to share the credit between two great engineers.

The obvious starting point for looking at the Ellesmere Canal is the line past Ellesmere itself, which leaves the Chester Canal at Hurleston Junction, just north of Nantwich. At first, the intention had been to make a connection with the Severn at Shrewsbury. It was clear from the start that there was no question of taking a very direct line through the aggressively hilly country, and in any case another important objective appeared. In 1796, a new Act was passed for the Montgomeryshire Canal that was heading south through Welshpool. This was to be joined by a branch of the Ellesmere, while at the same time a second branch would turn north towards the coal mines, limestone quarries and ironworks round Chirk and Ruabon. So the line was fixed, heading south to Welsh Frankton, where one arm headed off to the Montgomery. Although the southern sweep avoided some of the worst of the hills, life was never going to be easy for the engineers, and it was essential to keep a balance between shortening the distance to be travelled but at the expense of heavy engineering works, and making deviations with inevitable delays to traffic on the waterway. Jessop was a master of this art of creative compromise.

There is a bold beginning with a flight of four locks lifting the canal up from the Cheshire plain. The Chester Canal had been built with wide locks, but as it was clear that the Ellesmere Co. would ultimately be looking for connections with the established Midland system with its narrow locks, the decision to make this a narrow canal was inevitable. The Ellesmere is unusual in that it has no summit level as such, simply going uphill all the way to the Dee, which provided all the water that was needed, with some to spare. There is a reservoir at Hurleston which takes the excess water and distributes it as part of the general water system of the region. Without it, there might be no Llangollen Canal today. It was kept open long after commercial carrying ended, purely because of its water supply value. Then, when pleasure boating began, it enjoyed a new lease of life as one of the most beautiful and dramatic canals of the whole system. There are many enthusiasts – including the author – who can date a lifetime's fascination with waterways to the day they decided to take a boating holiday on the Llangollen.

At the top of those first four big steps up from the valley, there is a four-square house with neat bow windows standing by a typical bridge, enlivened by the use of alternating bands of red and blue bricks. The advantage of grouping the locks in a flight at the beginning now becomes clear. The canal has arrived at a rural landscape of hamlets and isolated farms, presenting no obstacle to progress and having no real need for a canal at their doorsteps, so the engineers could simply ignore them and let geography alone determine the line to be taken. Over the next six miles there are two locks close together at Swanley and three more at Baddiley, while elsewhere the canal maintains a reasonably direct line with just the odd swerve to keep the level. The only village of any size is Wrenbury and even that did not merit a diversion from the optimum line. Instead a wharf area was created well to the north of the old village centre, and around this a tiny canal settlement developed with mill, houses and a pub. Here, too, one can see one of the characteristic features of the canal, the lifting bridge. This is comparatively sophisticated. The bridge itself consists of a plain wooden platform, pivoting on one side of the canal and dropping into a stone-lined recess on the other. A tall wooden frame rises above the pivot point, supporting an overhead balance beam, with chains at either end. One set is attached to the far end of the platform, the other dangles down from the opposite end of the beam. By pulling on the latter, the bridge is easily raised – in theory, at least. Recent modification at some bridges has seen this simple system replaced by a windlass operating a hydraulic piston to raise and lower the platform, even though the superstructure has been retained. It has to be said that the new system at least removes one unfortunate problem that beset the old system - the absent chain.

The landscape ahead becomes notably hillier as the canal approaches the Welsh border but it turns away to the south, for an increasingly twisted course and with locks appearing more frequently along the way. The big leap comes at Grindley Brook. Up to this point, the canal has been making its way down a steadily narrowing valley, until finally it makes a sharp turn for a direct confrontation with the hillside. First come three conventional locks, closely followed by a three-lock staircase, which is itself a visual delight in spite of a less than romantic setting. There is a rhythmic pattern created by the familiar array of black and white balance beams and paddle gear, all emphasised by the repeated rows of ridges set into the lockside path to supply grip. The general rule of lock cottages being built in a simple vernacular is broken here. Telford obviously welcomed the opportuntity to practise his architectural skills, and turned a simple cottage into a cottage orné by adding a bow front, emphasised by a delicate, curved verandah. But the most important feature of Grindley Brook is the evidence it provides of Jessop's careful planning. There are twenty miles to go to the next lock on the way to Llangollen.

Now the canal squirms round a hill in a great U-bend. It is perhaps surprising to find such extravagant contour cutting in a canal of the 1790s, but as always with Jessop there is a rational explanation. The bend brings the canal close to the first of the major towns to be met along the way, Whitchurch. As at Wrenbury, the original plan was to skirt the edge of the built-up area, but as the local people proved enthusiastic supporters and users of the canal, a branch was added between 1809 and 1811. It closed again in 1944 and restoration was begun in 1989. A short section was reopened, but there are plans to carry it on to a new terminus in Jubilee Park at the heart of the town.

Having conquered the hills, the engineers now faced a more daunting challenge. When the glaciers of the last Ice Age retreated from the area, they left behind a series of scoured out

Thomas Telford's architectural skills on show at Grindley Brook lock cottage.

A feature of the Ellesmere Canal is the lift bridge. The overhead structure acts as a counterbalance to ease the work of raising the platform.

Above: 1. *Windmill End, Dudley No.2 Canal.* Below: 2. *Asylum lock, Hanwell, Grand Union Canal.*

Above: 3. *Lowsonford lock and lock cottage, Stratford Canal.*

Opposite Page
Above: 4. *Guillotine lock, King's Norton Junction, Stratford Canal.*
Below: 5. *Astwood bottom lock, Worcester & Birmingham.*

Above: 6. *Wolverton cast iron aqueduct, Grand Union.*

Left: 7. *The unusual lock gear on the Montgomery Canal.*

Right: 8. *Enclosed paddle gear on the Grand Union at Knowle.*
Below: 9. *Kensal Green gas holder, Grand Union.*

Above: 10. *Peartree Bridge, Milton Keynes, Grand Union Canal.*
Below: 11. *Talybont, Brecon & Abergavenny.*

Above: 12. *Wooded section on the Basingstoke near Fleet.*

Right: 13. *Crick Tunnel on the Leicester Arm of the Grand Union Canal.*

Above: 14. *The Boat Museum, Ellesmere Port*

Overleaf

Left-hand page, clockwise from top: 17. *Brynich aqueduct, Brecon & Abergavenny*. 18. *The inclined plane at Coalport*. 19. *Caen Hill flight at Devizes, Kennet & Avon*.

Right-hand page: 20. *The Rochdale Canal in Manchester*.

Right: 15. *Ellesmere basin, Ellesmere Canal.*
Below: 16. *The Pontcysyllte aqueduct.*

Above left: 17. Above right: 18, Below: 19, Opposite: 20

Opposite page:
Top: 24. *St Lawrence's Church, Hungerford, Kennet & Avon.*

Bottom: 25. *The Grantham Canal near Woolsthorpe.*

This page:
Left: 21. *Dundas aqueduct, Kennet & Avon.*
Below left: 22. *Lockside cobbles, Ancoats, Ashton Canal.*
Below right: *Ladies Bridge, Wilcot on the Kennet & Avon.*

Above left: 26. *Cotton mill near Dukinfield Junction*. Above right: 27. *The Huddersfield Canal and Saddleworth viaduct*. Below: 27. *Dereliction on the Dearne & Dove*.

Above: 29. *The Clyde Puffer,* VIC 32, *at Crinan.*
Below: 30. *Cobbled bridge over the Peak Forest, Dukinfield.*

Above left: 31. *Railway viaduct and Rochdale Canal, Todmorden*. Above right: 32. *Llangynidr, Brecon & Abergavenny*. Below: 33. *The summit of the Crinan Canal*.

hollows. Some filled with water to create the small lakes that dot the landscape, but others filled with vegetation. These marshes eventually dried out, and the vegetable matter rotted to create a black, oozing mass of peat. One of these lay right in the path of the canal, Whixall Moss. Before any construction could begin, a drainage channel had to be cut through the centre along the line of the canal. When the ground was firm enough, an embankment was piled on top of the peat, and the canal cut into that. Now it appears as a straight line, dividing reclaimed land of rich, peaty soil on one side from a wasteland of pools and scrub on the other. The latter now provides a rich habitat, particularly for insects.

At the end of the long straight is the junction with the so-called Prees Branch. Prees seems to have been unfortunate with its transport connections. The canal never got there at all, stopping two miles away at Quina Brook, and although when the railway came it got a station, that was a mile away as well. In its day, the Prees Branch was a valuable source of cargo, serving a series of lime kilns. It closed along with the Wrenbury Branch in 1944, but again has been partly reopened as far as a marina at Dobson's Bridge. The main line, having cleared the moss, was now confronted by The Meres, a sort of miniature Lake District, where the canal edges its way along beside the broader expanses of water. The largest of the lakes is known simply as The Mere and the town of Ellesmere stands on its bank. This was to become the most important point on the canal in its journey westward. A short arm leads down to the town wharf. Fitting neatly into the bend at the junction is Beech House, an L-shaped building that housed the company headquarters. Telford repeated the bow-front motif introduced at Grindley Brook, this time continuing the bow up through two-storeys to end in a semi-conical roof, producing a turret-like effect. Next to that is the maintenance yard with dry docks, workshops, forges and stores. The exposed timbers of the main building give an appearance which should be rather similar to that of the 'black and white' houses for which the area is famous, but ends up looking disconcertingly like a stockbroker Tudor house that has ended its days as a garage. The arm leads off to a town wharf with warehouse, hand crane and creamery. Ellesmere was a prosperous market town even before the canal arrived, and was famous for its cheeses. The rennet works are now over two centuries old, and it was this busy trade that made the town important enough to give a name to the canal.

Because those who travel this way today consider Llangollen to be the main objective, they assume it was always so. It was not: the main line continued south at Frankton Junction for a further eleven miles to Carreghofa and the junction with the Montgomery. What we call the Llangollen was originally the Pontcysyllte branch line. But rather than stay with the strict historical perspective, we shall do as most boaters do, and carry on towards Llangollen.

This route now seems delightfully rural, but in fact it passes through what was once an important coal mining region. The engineers, however, were looking further ahead towards what threatened to be extremely difficult crossings of two valleys, the Ceiriog and the Dee. There was much discussion over what course to follow, and before any final decision was reached Telford was whisked away for his brief spell of work completing the Shrewsbury Canal. As described earlier, he worked with William Reynolds to build a cast iron aqueduct across the Tern, and returned full of enthusiasm for the new material. A conventional masonry aqueduct over the Dee would have required immense supports to hold up the long, heavy stone trough and its equally heavy lining. A light iron trough was clearly a far better option. Jessop would have needed very little convincing. He was already in partnership with Benjamin Outram of the Butterley Iron Works, and Outram had recently constructed his own cast iron aqueduct at Derby. For many years, credit for the great aqueduct over the Dee was given exclusively to Telford – and often still is. It has to be remembered that Jessop was still the chief engineer, and the final decision was his. Who produced the detailed design for Pontcysyllte? Unless new evidence appears, we shall probably never know. And that is why it seems only fair to give credit to both Jessop and Telford. In any case, what really matters is the structure itself, one of the glories of the British canal system and quite grand enough for the honours to be shared. Once the decision had been taken to cross the Dee at the level of the high aqueduct, all the rest of the pattern fell into place.

There were to be two locks at New Merton to bring the canal up to the required height, but having arrived, a large amount of engineering works would prove necessary to keep it there. The approach to the Ceiriog valley hugs the hillside of Chirk Bank, heavily reinforced on the downhill side to prevent slippage. The canal swings west along the river valley until the chosen crossing point is reached, then makes a sharp turn towards the aqueduct itself. At first glance, this seems to be a conventional masonry aqueduct, carried on ten arches. But the depth of masonry above the arches is considerably less than one would find in earlier stone structures. The secret is hidden from view. In the older works, the actual channel for the water was constructed in stone, made watertight by puddled clay, all adding up to a considerable weight to be supported. To provide the necessary solidity, the spandrels would have been filled with rubble or clay. Here the bottom of the trough is covered by interlocking, flanged iron plates, and the sides of the trough are made of ashlar, carefully dressed stone, backed by hard bricks set in waterproof cement. The spandrels are hollow, braced by longitudinal walls. The metal plates, as well as providing a waterproof lining, also form a continuous tie beam, adding to the overall strength. So less masonry was needed in the arches, and the whole structure has a light, airy appearance. The only thing that detracts from this very fine aqueduct is the fact that it is literally overshadowed by the later railway viaduct, built right alongside.

Crisp black and white provide a pleasing effect on the stables at Frankton Junction, where the Ellesmere and Montgomery Canals meet.

Above: *A boat in the iron trough of Pontcysyllte aqueduct.* Below: *Pontcysyllte, aerial view.*

Once across, the canal vanishes into the 459 yard long Chirk Tunnel which, unusually, enjoys the luxury of a towpath. Chirk was an important connection point, for it was here that it was joined by a tramway from the slate mines and woollen mills of Glyn Ceiriog. A deep cutting now carries the route forward, where there is a repeat performance of the approach to Chirk, with the canal first swinging round into the Dee valley, followed by a sharp turn to Pontcysyllte. That this is a remarkable structure scarcely needs to be said, but it is even more interesting than it appears at first. Perhaps the most remarkable thing of all is that it followed on from such a tentative beginning, the low, short aqueduct at Longdon. Here we have an aqueduct which is 1,007ft long and rising to a height of 121ft above the Dee on nineteen arches. Again, some of the innovation is hidden from view. The piers are built in the usual way, of solid masonry, up to a height of 70ft, but above that they are hollow, braced by cross walls. The iron trough seems, by contrast, to be obviously of revolutionary design, but even here things are not quite what they seem. It has to be remembered that there was no one with any experience of using iron in this way, and Telford had been trained as a stone mason. From his youth he had thought of bridges in terms of arches, with *voussoirs* of wedge-shaped stones pressed together to provide the necessary strength. And that is what we find here, wedge-shaped iron plates bolted together. The individual plates come in a variety of shapes and sizes, meshing together in a complex jigsaw, and as a result far more patterns for casting had to be prepared than were strictly necessary. In a belt-and-braces approach, added security is provided by cast iron arches, again of individually cast wedge-shaped pieces, running between the stone piers. But considering just how revolutionary the structure was, the engineers can be forgiven for exercising caution. One successful innovation was to add a towpath within the trough, cantilevered out over the waterway. Although the gap between the towpath and the edge of the trough is scarcely wider than the beam of a boat, the water extends under the towpath and makes for a free flow. It was a structure meant to last, but it would obviously need maintenance at regular intervals, so a plug was set into the middle of the trough. With stop planks in place, the plug can be pulled and the whole trough emptied in a spectacular cascade.

The final stage of the journey is again more complex than it seems. The canal went straight on from the end of the aqueduct to Ruabon. The original plans had called for a continuation to the Dee, but it was wisely decided that the expense of hacking a way through the hills would never be reclaimed in revenue from the moribund river. On the other hand, Ruabon was at the heart of a complex of ironworks, including Plas Kynaston, where the metal for Pontcysyllte was cast. Telford was to use the works again to supply castings for bridges in Scotland, and the parts would start their journey to the Mersey by crossing the iron aqueduct. Once again, this once important route has been reduced to a stump, used only for moorings. The remainder of the route westwards consists of the navigable feeder to the Horseshoe Falls. It hugs the valley side, twisting to follow every curve in the land, as it nears the upper reaches of the river. Along the way, however, there is ample evidence that this branch also had its share of valuable trade. A double arched bridge crosses both the canal and a seemingly featureless track. The latter marks the line of a tramway which connected to limestone quarries high up in the hills. At Llangollen itself there is an extensive wharf and wharf building, which is the centre from which horse-drawn boats take passengers for trips on the canal. This is no modern innovation: the pleasure boats have been plying their trade for over a century. A short way beyond that, the navigation ends and the canal has its beginning in the sluices controlling the water from the Dee.

At the opposite end of this canal, the route was completed from Chester to the Mersey at Ellesmere Port. This was a straightforward connection, which acquired new significance – and its most important structures – at a later date, with the building of the Birmingham & Liverpool Junction Canal, which will be dealt with in the final volume. It is now time to turn to the rest of the Ellesmere and its connection with the Montgomeryshire. It is worth remembering that what we are now going to be looking at was the original main line, so that this was an important junction. But a serious breach occurred near Frankton in 1936, at which time traffic was so low that no one thought it worth putting in the money to repair it. That effectively finished off the canal, though it was not officially closed until 1944. Although a movement towards restoration got under way in 1968, large sections of the canal remain unnavigable at the time of writing. However, there is still a great deal to be seen even on those sections which are no longer in water.

The story of the building of the canal is complex. The original objective of the Mongomeryshire – not the Montgomery, for it was never intended to go to the town – was to link the 'Porthywain Lime Rocks' in Shropshire to somewhere 'to or near Newtown'. It was then decided to link in with the Ellesmere near the quarries at Llanymynech, and to add an arm to serve quarries and coal mines round the village of Guilsfield. Work began under the engineer John Dadford, but came to a halt seven miles short of Newtown at Garthmyl. A new Act of 1815 complicated matters by creating two companies. The existing route was now handed over to the Eastern Branch of the Montgomeryshire, and a second, separately funded company, the Western Branch was given the job of completing the line to Newtown. This time the engineer was William Jessop's son, Josias. It has to be said that work on the line, particularly in building the aqueducts, was less than satisfactory, so a third engineer, George W. Buck, was appointed to sort out the problems. He started on the Eastern Branch in 1818 and moved on to the Western Branch in 1832. Further remedial work was carried out by his successor, J.A.S. Sword. So when we look at the canal today, we are seeing the workmanship of three generations of engineers, not to mention that of the modern restorers.

The story of the Ellesmere section is also complicated. The original intention had been to create a main line from Frankton to the Severn at Shrewsbury, with a branch to the same productive quarries that were the principal objective of the Montgomeryshire. In the event, the Shrewsbury line never got further than Weston Lullingfields, and was reduced to branch line status. The old branch line to the quarries was now the important route, which gained in status by being linked to the Montgomeryshire at Carreghofa Locks.

Starting with Frankton Junction and the Ellesmere, Jessop was able to take a reasonably direct line towards the limestone hills. The route was downhill all the way, so there were no water problems on the canal, which like the rest of the Ellesmere system was fed by the Dee. There was an added financial bonus once the junction was made, as they could charge the Montgomeryshire for the use of their water. There is a busy beginning, with the canal at once dropping down through a double lock, overlooked by a neat, single-storey toll house, quite unpretentious in style, but made to look dramatic by the use of black paint on the stone blocks of the quoins, which stand out against the white walls. There are two more locks, with lock cottage, stables and a former pub to complete the group. It is rather reminiscent of Hurleston Junction, where once again locks are crowded close together at the start, giving a total fall of 31ft. The boatyard between Locks 3 and 4 has a certain historical significance, for among the vessels built here was a certain Shropshire fly – a narrow boat, that worked day and night, using relays of crews and horses. It was called *Cressy* and was to find fame as the

The navigable feeder that links the main line of the Ellesmere Canal to Llangollen, with Castell Dinas Bran in the background.

Narrow Boat in the title of L.T.C. Rolt's classic book. The other feature of importance here is the derelict arm that leads away for a much shortened journey of just a few yards, rather than the few miles, to the original terminus at Weston Wharf.

Restoration is very far from complete, but the fight began back in 1969, when a 'Big Dig' resulted in the clearing out of the canal right through Welshpool. Some 200 volunteers turned up and the result was the formation of the Waterways Recovery Group under the leadership of bustling, bearded Graham Palmer. Sadly Graham died while still comparatively young, but his work is remembered at the next lock, which is named after him. It is hard to think of a more appropriate memorial – apart from the completion of the work to which he devoted so much enthusiasm and energy. Now the engineers faced their first major obstacle, the crossing of the River Perry on a conventional three-arched masonry aqueduct The next very interesting site is at Rednal, where a roving bridge carries the towpath from one side of the canal to the other by curling back in on itself. Here too is a basin, built for the interchange of goods between the canal and the Shrewsbury & Chester Railway which arrived in 1848. There are also the remains of a fertiliser factory and an interesting warehouse with an exposed wooden frame. It is slightly odd in that the door by the towpath is not wide enough for bulky goods, nor is there any indication of a loading bay or hoist above it. The likeliest explanation is that this was used by passengers on the Swift Boat Packet Co. service between here and Newtown.

The next section has curiosity value. The story is that a local vicar requested a diversion to pass near his house, and as he was a shareholder the company agreed. Then, after work had begun, he changed his mind. Having caused so much nuisance, he passed over a parcel of land as compensation, which enabled the company to build a dead straight section of canal with a road alongside. It ends at Queen's Head, with an extensive wharf, warehousing and two hand cranes. A tunnel under the road carried a tramway bringing sand to the wharf.

The canal is still heading downhill through Aston locks to arrive at Maesbury Marsh and another splendid collection of canal buildings, including the local equivalent of Beech House at Ellesmere, a smaller version this time of the main offices and known as Sycamore House. The nearby, three-storey Navigation Inn is quite unusual with an attached warehouse. The latter can be seen to have started off as a single storey stone building, later extended upwards in brick. This was a very important location, as a short arm a little further on leads off to the great complex of Maesbury Hall Mill, where corn has been ground since the 1840s, going through the full power range, starting with water, going on to steam and ending up with electricity.

Now, as the hills get ever closer, tramway connections become increasingly important – a colliery tramway at Gronwen, a quarry tramway at Crickheath. The canal itself is forced into an increasingly convoluted course as it approaches Llanymynach, overlooked by its shattered hill. There have been quarries here since Roman times, when they were worked for lead, copper and zinc, but it was limestone that was quarried in the canal age, mainly for burning into lime for agriculture. Everywhere on the hillside are the scars of craters and cliffs, and the line of the canal tramway can still be traced with, in one place, an old broken brake drum still standing above an incline. There are also, as at many places along the canal, the remains of lime kilns, notably at Pant.

The end of the Ellesmere line is now near, but there is one late addition to the route. When the West Shropshire Mineral Railway was built in the 1860s, a new aqueduct had to be constructed across the tracks. The line of the canal had to be altered, moved a little to the south, and the aqueduct used the materials of the new age, wrought iron instead of cast iron for the trough, which is carried on cast iron pillars. The locks at Carreghofa mark the division between the two canal companies, and differences appear immediately. The most obvious newcomer is the unique cast iron paddle gear, devised by Buck. In place of the conventional upright rack and pinion lifting ground paddles vertically, what appears at the lockside is a triangular frame, with circular gear wheels engaging with a pivoting, toothed quadrant. As the quadrant moves, the motion is taken via a crank to a horizontal plate that slides backwards and forwards across the sluice in the floor of the lock chamber. The parts were cast at the famous Coalbrookdale iron works. It is ingenious, certainly, but it is difficult to see why it was necessary to replace a simple mechanism which had worked well for decades, with a more complex one. This being a junction, there is the inevitable cluster of buildings, including lock cottage, toll house and toll-keeper's house. They are not all of the same date. In the early years one man was both toll collector and lock-keeper. He was expected to be ever watchful of traffic movements, so the bow front to his cottage is not there for decoration, but to ensure a clear view up and down the canal. In the event, the work proved too much for one man, so it was divided and a second cottage built around 1820. The Montgomeryshire company were reluctant to rely on the Ellesmere for their water supplies. An old watermill near the canal was already served by a leat from the River Tanat, and this was enlarged to serve as a canal feeder. Then in the 1820s, a more substantial weir was built across the river, and the feeder enlarged to take boats, providing an independent route to the quarries and, most importantly, one free of Ellesmere tolls.

There was never much question about the overall line of the canal as it headed off to follow the Severn valley, but there were some difficulties to overcome, with crossings of the flood plain of the Vyrnwy and the shallow valley of the Bele Brook, both of which called for massive earthworks. Five aqueducts were to be built on the line, and all were to give problems over the years – problems that were to be solved in an interesting variety of ways. The first of these major structures appears right away, the formidable Vyrnwy embankment. It stands up to 20ft high, and spreads at its base to a width of 100ft. Along its length there are a number of arches to let flood water though. It runs for nearly a quarter of a mile and ends at the Vyrnwy aqueduct. There is nothing unusual about it, a perfectly normal stone aqueduct carried on five arches. The actual trough is stone, backed by puddled clay and encased in masonry, so that it appears very sturdy. Appearances are deceptive. It was not long before one arch had collapsed and needed repair, and remedial work went on for years. You can still see the evidence, the ends of iron tie beams are dotted all over the spandrels, and a complex pattern of exterior ties holds arches together. Iron appears again at Bridge 97, this time as planned, not as a remedy for old errors. The flat platform rests on fish-bellied beams, which have one side curved to provide greater width and greater strength at the mid-point between the supports. Here is another wharf, with the very handsome salt warehouse, built into a slope, so that at the upper level it appears single-storey and carts can unload directly into the upper floor, while at the canalside it rises to its full two storeys.

The next important settlement is at Canal House or Clafton Bridge. Two large basins were excavated and the material was used for the Vyrnwy bank. Here once again, warehouses were built against the rising ground to provide access at different levels. There are canal workers' houses and Canal House itself, which belonged to the owner of nearby lime kilns. The kilns

Vyrnwy aqueduct on the Montgomery Canal. The array of tie beams show that structural weaknesses have developed over the years.

The Montgomery Canal Company timber-framed warehouse at Rednal. The bridge sweeps up to the road, where there is access at first floor level.

are regular features alongside the canal and wherever they appear there are bound to be wharves to take away the lime. Some of the best preserved are at Maerdy. In general they are built so that the top of the kiln is at canal level, so that the heavy loads, the limestone itself and the coal for fuel, can be brought in by canal and tipped in, while the lime is removed at the bottom.

A junction appears at Burgedin, where the Guilsfield Branch leads off towards the hills. It was short, but quite important enough to warrant a deep cutting, which was so impressive at the time that the nearest hamlet took the name of Deepcutting. It was used to bring both coal and limestone to the main line, which now drops down again through two locks. There is an unusually elaborate lock cottage integrated with warehousing space, stabling and office and with a feature that is surprisingly common on this canal, a neatly built pig sty. These locks lead down to the second big embankment, which though not as high as Vyrnwy is three times as long. Beside it are the 'borrow pits' from which the earth for building was excavated, which were later flooded and planted with willow for basket weaving.

Now the canal's downward journey ends and it begins to climb for the last leg of the journey, through a staggered array of locks. The first settlement, Pool Quay, is unusual. Here the canal comes very close to the Severn which, rather amazingly, was navigable this far upstream in the eighteenth century. So the settlement here, or what remains of it, is orientated as much towards the river as towards the canal. More importantly, the line is now approaching Welshpool, past a very imposing array of limekilns at Buttington. At the edge of town, the Lledan Brook is crossed on a short aqueduct. This is a very late addition, not even Buck's work but that of his successor, J.B.S. Sword. It has a cast iron trough, with a towpath to either side supported on stone arches. The overspill weir at the end of the aqueduct took water away to a mill leat. This is the first big town on the whole route, and in its day it was an important centre for flannel manufacture. So we find a maintenance yard which has changed a good deal over the years. Among the buildings is a covered-over saw pit, but much of what we see today only dates back to the days when the canal was taken over by the Shropshire Union Railways & Canal Co. They were not much given to architectural extravaganzas, and preferred cheap timber sheds with corrugated iron roofs, though they did add an internal railway system for greater efficiency. The most imposing wharf building is a nineteenth century warehouse with a hipped roof that has been extended over the wharf to form a canopy above the loading area. The canalside cottages were part domestic, part warehousing.

Leaving Welshpool, the canal arrives at Belan locks and the most imposing of all the canalside banks of kilns. The slow climb continues now towards Newtown to the aqueduct over the Luggy Brook. Cast iron again here, but designed by Buck, and clearly the pattern for Sword's later work. The next aqueduct, however, is a much grander affair, crossing the River Rhiw at Berriew. This started off as a masonry aqueduct, but once again reinforcement was needed, this time supplied by an outward casing of hard, blue engineering brick. Two tones were used to give a decorative, banded effect. The next aqueduct at Aberbechan is masonry, with three low arches separated by round pillars to act as cutwaters. This was the work of Josias Jessop, who seems to have fared no better than anyone else, for once again it has been heavily patched in brick. Jessop's other main contribution to the canal is not so immediately apparent. A feeder joins the canal from the River Severn at Penarth, and it is there that the real work went on with the construction of two immense curving weirs across the river.

The end of the line at Newtown must have been an impressive sight with a terminal basin, banks of lime kilns surrounded by heaps of coal and stone, a surrounding road system, warehouses and workers cottages. Some of the latter survive in Dolafon Road, but little else – no basin, not even a canal. This was and is the end of the North Wales canal system which remained isolated from its southern neighbours.

5. South Wales

The canals of South Wales have been even less fortunate than those in the north, and at one time it seemed that they might all be lost. The most extensive system was made up of the Monmouthshire Canal and the Brecon & Abergavenny, which met in an end-to-end junction. But traffic steadily dwindled to the point when in 1933 not a single toll was paid. The Monmouthshire was abandoned, but the Brecon & Abergavenny was kept open because it still had value as a water supply system – but that did not mean it had to remain navigable. At least the main canal remained intact into the age when pleasure boating took over from cargo carrying. It was recognised as one of the most beautiful of all Britain's canals, following what had now become a purely rural route through the Brecon Beacons National Park. It was transformed from a semi-derelict waterway to a fully navigable canal in 1970. Other canals in the region have been less fortunate and though there are ambitious restoration schemes in hand, some it seems have gone for ever, along with the industrial world they once served.

It might seem sensible to begin with the canal that is actually open, but the Brecon would never exist without the Monmouthshire. That received its Act in 1792, the Brecon a year later. The aim of both canals was the same, to provide a route to the sea via the navigable River Usk at Newport. In a sense, the Brecon was no more than a further extension of the older canal, and both carried the same basic cargoes – coal and iron. The Monmouthshire had travelled no more than a mile and a half from Newport when it divided: the main line heading north towards Pontypool, a branch heading up Ebbw Vale to Crumlin. This was not easy country for canal builders. The valleys rise steeply, needing thirty-one locks in the nine-mile-long main line and a further thirty-two in the eleven mile Crumlin branch, but the effort proved well worth while. By the 1840s the canal was carrying almost 250,000 tons of iron and over 600,000 tons of coal a year. This is very impressive, but in the success story are hidden the origins of ultimate failure.

The boats of the South Wales canals were unlike those of the English system. They were approximately 60ft long by 8ft 6in beam, but were double-ended and because the distances travelled were generally short, they had no cabin. They normally loaded up to twenty tons. So the iron and coal trade alone accounted for over 40,000 boat movements a year. With so many locks, congestion was inevitable, and as early as 1830, right at the beginning of the railway age, there was talk of turning to steam locomotives in place of horse-drawn boats. The Monmouthshire Canal Co. was transformed into the Monmouthshire Railway & Canal Co., and in 1852 a line was opened close to the old canal, joining Newport to Pontypool, soon extended by the Taff Vale Extension. Railway lines not only ran alongside the older canal, but vaulted arrogantly over them, nowhere more spectacularly than at Crumlin. Here the 200ft-high iron trestle viaduct, with its open lattice piers seemed almost too fragile to carry anything, let alone a steam train. It was one of the great engineering marvels of the day, loudly proclaiming that a new transport era had arrived, and the old canal was doomed to dwindle and die.

There is still a good deal to see on the old main line to Crumlin, and quite enough to show that the engineer, Thomas Dadford Junior, was very much his own man. This is particularly noticeable in his arrangement of locks, and the best place to see this is Rogerstone, starting at The Fourteen Locks Canal Centre. Here the locks have been cleaned out, though cascaded, and the basin and side ponds cleared. The Centre itself stands by the wide basin, which also serves as the first of a series of mini-reservoirs for the flight. Any excess water is channelled away down a well-constructed stone culvert to another pond lower down, which in turn feeds into the locks, and so on down the flight. The locks themselves are unusual in that they appear at first glance to consist of pairs of double locks or risers, but they are actually separated by very short pounds, so that they do not share the same gates as they would in a conventional staircase. It seems odd, rather reminiscent of Brindley's near-staircase at The Bratch on the Staffs & Worcester. It does, however, help with the complex system of water distribution.

Elsewhere, the canal does occasional disappearing acts under more recent developments, but where there are long sections in water, one can see another aspect of Dadford's planning. Moving on towards Crumlin, there is a section still in water at Risca, where the canal is cut along the contours of the hill, enjoying an airy location high above the towns clustered down in the valley floor. On a canal such as this, which has been disused for a long time, it is always difficult to decide what was part of the original plan and what is the result of neglect. When a canal is built with a very severe gradient, water supply is clearly a major consideration, and any source of water to feed the numerous locks is going to be welcome. What is now very noticeable is that hill streams appear to have been allowed to run straight into the canal, with no traps to catch and hold back the silt. In busier times, the constant passage of boats and the flow of water down the locks would have had a scouring action, but it must always have been rather troublesome. Everything one sees, speaks of a well-designed and well-constructed canal, not least the neat stone bridges that dot the route. It is also, it has to be said, very attractive, particularly in the upper reaches where there are wide views out over the hills. It cannot, however, quite match the next waterway on the list.

The Brecon & Abergavenny Canal began life as the Brecknock & Abergavenny in 1793, and it retained its independent existence until the situation was rationalised in 1865 when the Monmouthshire bought up its old neighbour. It was a short lived ownership, for in 1880 it was all taken over by the Great Western Railway and yet another name was added to the list: the Monmouthshire & Brecon Canal, confirming in words what had always been true in practice; that they formed one continuous route. If this suggests that there was always a steady, amicable relationship, then this was not altogether the case. Both companies had extensive tramway connections, and they were not above trying to use these to poach traffic from the other. There is probably no happier hunting ground for the tramway enthusiast than the Brecon & Abergavenny which in its hey day had fifteen separate tramway connections, and there is still a good deal of physical evidence for their presence. If the Brecon makes little sense without the Monmouthshire, it is equally true that it is meaningless without its tramways. Those who come this way today, by foot or by boat, will enjoy a superbly scenic canal, one of the best in Britain, but might well be left wondering why it was ever built. There is nothing in the immediate surroundings to even hint at likely cargoes. The only way to answer the riddle is to follow not just the canal but the tramways as well.

There can never have been much question about the best route to follow to reach Brecon, nor about the engineer. The job went to the man who had proved his worth on the Monmouthshire, Thomas Dadford. At some stage he would have had to leave the Afon Lwyd valley and reach the Usk. Abergavenny featured in the name of the canal from the first, but

The toll house at Pontymoile marks the point where the Brecon & Abergavenny made a head-on junction with the Monmouthshire Canal.

if he had taken the canal to the town, he would have had to lock down a long way to reach the Usk, then build more locks up the Usk valley. Dadford's far more sensible solution was to ignore Abergavenny altogether. Instead, he took his canal along a contour round the hill of the Blorenge, high above the Usk, and let the valley rise to meet him. The result is an extraordinary example of contour cutting on a difficult hillside route that continues for twenty-three miles until the first lock is reached. The outstanding beauty of the scenery along the route, combined with having just six locks to negotiate in over thirty-three miles, makes this an ideal canal for holiday boating.

From the basin that marks the junction of the two canals at Pontymoile, the canal narrows down so that the boats could be gauged and appropriate fees collected at the toll house. This is an interesting little building with a curved end wall facing the canal, and the curve is carried on up into the slate roof. Like many of the canal buildings along the way, much of its charm derives from the contrast between crisp whitewashed walls and the changing shades of grey in the slates. Then the canal sets off for its journey round the hills with a short hop over the river on a simple aqueduct. The line is now entirely determined by the shapely folds of the hill, which sometimes force the canal into extraordinary convolutions. At Mamhilad, although the canal is supposed to be heading south, it is forced into heading in the opposite direction for a while before another turn sets it back on course. The most extravagant bend of all takes the canal round the two sides of a deep cleft in what is known as the Ochram Turn. This extreme method of contour cutting on a steep hillside created its own problems. It was not just a question of cutting a channel into the slope, but the excavated material had to be used to create a level terrace for the towpath, and unless this was well consolidated, there was a danger of landslips. Breaches have, in fact, been a recurring problem.

The Anatomy of Canals

If Dadford was unwilling to leave his airy line even to visit Abergavenny, he was never going to make diversions to reach smaller towns and villages, so extensive wharves had to be built at the nearest approach. At Goytre, there are signs of important canal traffic, with lime kilns sitting in among the more familiar canal warehouses. An even more interesting group appears a little further along the way at Llanfoist. The first clues to the nature of this site come with two canal crossings, first by a flat-topped bridge over the water, the second a tunnel under the canal. Why two? The answer is the tramway system. The bridge carried the trucks, and pedestrians wisely kept well clear by using the tunnel. The wharf itself is dominated by two buildings. The first is a three-storey house, very substantial, but somewhat altered over the years. The second is the warehouse, built alongside the wharf, which boasts a small crane. A short arm takes boats in under the warehouse floor for loading. The upper storey is supported on pillars at the water's edge, and stands up against the hillside at the other side. Here trucks from the tramway had direct access to the upper level and grooves on the floor, to the right gauge, indicate where they could be wheeled in above the waiting boats. What is not so immediately obvious is the nature of the tramway and where it came from. There is a clue in the wharf manager's house, where a window in the end wall gives a clear view out to a footpath rising up the steeply wooded hillside. This was the incline which carried trucks down the slope of the Blorenge hill in two stages. Anyone who walks up the path will find the old stone sleeper blocks into which the rails were spiked. There is a break half-way up, a platform which held a brake drum. Although horses were used on the level sections of the tramway, movement up and down the slope was by cable haulage. At the top of the hill, the tramway divides. One branch went up to limestone quarries, and the stone was carried down, past the wharf and over the flat-topped bridge to kilns in Llanfoist itself. The second branch leads up to the

The Brecon & Abergavenny Canal hugs the hillside, keeping to one level as it sweeps round the slope of the Blorenge hill.

Wharf, warehouse and wharfinger's house at Llanfoist. It was here that the canal was joined by Hill's tramway bringing iron from the furnaces at Blaenavon.

Blaenavon iron works, established in 1789. The remains can still be seen, including an impressive array of blast furnaces and a water balance tower, used for moving material between the upper and lower levels of the site. The iron works and the nearby colliery – now the Big Pit Mining Museum – kept the canal busy, and what is now a hire boat base in a tranquil canal setting was once a vital link in a major industrial transport system.

There are similar stories to be told at other wharves, with an interesting variation at Llangattock, where a tramway was built from Bailey's iron works at Nantyglo. It curved round the slope of Myndd Llangatwg, passing the quarries that provide a hard, straight edge to the northern face of the hill. At the wharf, the tramway emerged above the big battery of lime kilns, so that the stone could be loaded straight into the top. From here the canal bends to run close by the River Usk, and the long pound comes to an end as the valley steepens towards Brecon. There is one isolated lock, then the canal turns sharply to cross a tributary, and climbs again through a further four locks.

The valley floor is broad, but the hills still loom high to the west of the line, and one ridge extends right down to the riverbank. The only way forward was by tunnelling, but there was no extravagance, just a low tunnel with no towpath, a featureless hole in the hill. There is a cutting at the far end leading to Talybont wharf and a modest aqueduct over the Caerfanell, bustling down from the hills, where it has been dammed to create the Talybont reservoir. There is one more major obstacle to be overcome. The hills are still crowding in on the west, but across the far side of the Usk the going is much easier. It made a lot of sense to avoid digging tunnels and cuttings by constructing an aqueduct. Brynich aqueduct is in a fine situation, and when seen from the riverbank could easily be mistaken for a road bridge. Stone buttresses separate the four low arches and sharp cutwaters thrust out into the foaming Usk. Once across, the canal takes a sharp bend to continue along the line of the river, and there is one last lock before the final run into the outskirts of Brecon. The canal ends at a new basin, but once it extended into the centre of the town.

The story of the next two connected canals is essentially similar to that of the previous two: the Glamorganshire and the Aberdare Canals both served iron works, were connected to each other and to a port. The Glamorgan linked the great iron-making town of Merthyr Tydfil with Cardiff, while the other ran from this canal at Abercynon to Aberdare. The Glamorganshire Canal had to struggle up the valleys with fifty locks in the 24½ mile journey, ending up 543ft above sea level. It prospered for a while, but eventually succumbed to competition from the Taff Vale Railway. The northern section closed in 1898, from Abercynon to Pontypridd in 1915 and the rest in 1942. The Aberdare was no more fortunate. Over the years, whole sections have been built over, locks have been filled in and buildings demolished. There are still a few remains to be found. The old terminus basin at Aberdare still exists, with a wharf house and the stump of an old crane. There is a reedy trickle leading away from it, but that soon vanishes again under a new road. At the far end, the Navigation Inn remains by the site of the old wharf, with a sign optimistically showing pleasure boats moored by the pub. The wharf itself, however, has gone and in its place there is now the local fire station. Here, too, is a plaque marking an event that was to have a decisive influence on the world of canals, and in particular on the early closure of the northern Glamorganshire. The upper end was the most heavily locked, and there were serious delays as boats waited their turn. The iron master, Richard Crawshay, had a controlling interest and demanded precedence for his boats, so his rival Samuel Homfray and others promoted a separate transport route, a tramway from Abercynon to the Homfray works at Penydarren on the edge of Merthyr Tydfil. It was here in 1804 that the Cornish engineer, Richard Trevithick, arrived with his latest invention, a steam locomotive, that successfully hauled a train of trucks up and down the tramway. This is the event recorded on the plaque – the first public demonstration of the steam railway. The old tramway can be walked, and lines of stone sleeper blocks can be seen in place near Quaker's Yard, while alongside it is the railway that was to prove fatal to the canal.

Parts of the Glamorganshire Canal do exist. In Merthyr Tydfil, a section of the canal has been excavated outside iron workers' cottages in Chapel Row. The remains of a pair of canal boats have been preserved here, together with a cast iron bridge which originally crossed the canal at Rhydycar, about a mile away. At the far end, a section can be seen in the Cardiff suburbs in what is now the Glamorganshire Canal Nature Reserve, and there is one strange reminder of the past right in the heart of the city. To reach the docks, the canal had to pass under Queen Street in a tunnel, which is now used as a pedestrian underpass. Its old function is remembered by lock gear set at one end – and this must be the only urban underpass to boast a raised towpath.

Swansea was the other port to be served by canals, the Swansea Canal itself and the linked Neath and Tennant Canals. These in turn were joined to the Cardiff system by a tramway from the terminus of the Neath Canal to Aberdare. The Neath Canal was first on the scene, with an Act of 1791. It followed a very obvious route from Glyn Neath in the north along the line of the river, past Neath itself to Briton Ferry and quays alongside the navigable River Neath. Today, the latter area is rather sadly forlorn, the quays lost from view under new devel-

One of the surviving sections of the Neath and Tennant Canal still in water at Resolven.

opments and standing in the shadow of the main road and motorway viaducts. The canal itself has been reduced to little more than a ditch. From here there is a steep but steady climb, with nineteen locks in the thirteen mile journey. The Tennant Canal is unusual in that it was a private waterway, rivalling the Duke of Bridgewater's in size and importance. George Tennant established docks on the east bank of the River Tawe, across the water from Swansea, to link with his collieries near Neath. It was obviously a sensible move to make a junction with the Tennant, so he extended his canal to Aberdulais, where an aqueduct was built across the river. The structure is low, carried on ten semicircular arches, with the riverbed beneath the arches paved to reduce turbulence round the stonework of the aqueduct. Immediately upstream of the aqueduct is a weir, and a set of sluices allows river water to feed into the canal by the lock at the southern end of the aqueduct. It seems to be the fate of the more important features on this canal system to be overshadowed by later works. This time it is a tall, five arch railway viaduct that crosses the river beside the canal. The canal junction itself was completed in 1824. In the meantime the Swansea Canal had been constructed, which also contained an interesting feature. A short section, just a mile and a half, passed through the Duke of Beaufort's estate near Morris Town, now Morriston. The Duke paid for the section himself – and collected the tolls. The canal followed the line of the Tawe to Abercraf, where a tramway was built up to the stone quarries in the hills.

These canals of the Welsh valleys are full of interest and diversity and well repay time spent on investigation. It can, however, be a very exasperating affair: it sometimes seems that just when things are getting really interesting, the old canals are lost beneath modern buildings or roads. It seems unlikely that there will be full scale restoration of all these waterways, but sites such as the Rogerstone centre show just how valuable a project that stops short of a return to navigation can be.

6. Southern England

The most ambitious of the new schemes was for a canal that would link the two old river navigations, the Kennet and the Avon, to provide a through route from Bristol to the Thames and so on to London. It was planned as a majestic waterway with locks 73ft long and 14ft wide, to accommodate both narrow boats and river barges. The engineer who set to work in 1794 was John Rennie, and he sent his plans to William Jessop, who suggested a number of improvements. The end result was a canal that boasted two magnificent aqueducts, a tunnel and what, if not the longest, is certainly the most imposing flight of locks in Britain. All of this cost a great deal of money, which was not always readily available as estimates were constantly revised – and never downwards. In the event, the canal was not fully open until the end of 1810. As with so many waterways, early prosperity was succeeded by years of struggle, and boatmen on the canal could see all too clearly where the problem lay. The Great Western Railway paralleled the canal for much of the way, and the engine drivers could literally look down on the barges and their steady, plodding horses. In time, the GWR took over the canal, and although they kept to their statutory duty to keep the waterway open, they did nothing whatsoever to encourage trade, and maintenance was kept to an absolute minimum. The decline led to inevitable closure, but thanks to the enthusiasm of volunteers, at first doing the work themselves and then fundraising for the professionals, the canal enjoyed a grand second opening in 1990.

The difficulties faced by Rennie in his fifty-seven-mile-long journey were particularly troublesome at the western end. First there came a steep climb up from the River Avon at Bath, which he tackled head-on by building the seven lock Widcombe flight with a total rise of over 60ft. During the long period of restoration, a road was built across the locks near the bottom of the flight, and the restorers overcame the problem by leaving one lock out and doubling the depth of the survivor, creating the fearsome deep lock, with a rise of nearly 20ft. There was inevitably a serious problem in supplying water for the flight, so the company acquired the old Thimble Mill site and installed a steam pumping engine to feed water back up to the top of the locks. Now the pump has gone and Thimble Mill has reappeared, this time as a restaurant and hotel. At the top, there is a lock cottage, which has a quiet Regency charm, altogether appropriate for its setting.

Rennie had more than just engineering problems to contend with. Bath was one of the most fashionable and elegant cities in Britain, and the natural line of the canal would take it right by Sydney Gardens, where ladies and gentlemen paraded in their finery. Local opinion was quite clear: the cream of society did not wish to have their strolls disturbed by the sight of dirty, ugly coal barges, not to mention dirty, ugly bargemen. So the canal was tucked away in two short tunnels, and what was visible was given a suitable refined treatment. One tunnel portal is decked out with stone swags, and a fiercely bearded head peers out through the petrified curtains. Bridges are delicately ornate structures in iron, with roundels between the arches and topped by slender railings. The second tunnel is topped by a splendid house, where proportion is all. It is classic Georgian, with tall round-headed windows on the ground floor, and plainer sash windows above the string course. This was no private home, however, but the company offices.

Now the canal heads off towards the country in an airy situation high above the Avon valley, with a bird's-eye view out over Bath. Then as the river turns south, the canal swings away with it, passing Bathampton. Here is an unusual canalside pub, built into the canal bank, so that access from the towpath brings visitors in at first floor level. This is a long pound, nearly ten miles in all, and in spite of the pump-back system at Bath, water supply was always a problem. There was water in plenty and clearly in sight down in the Avon. The question was how to raise it. At Claverton, there was already an immense weir across the river, with water diverted down a leat to a mill. The canal company bought the mill and converted it into a pumping station. No need to buy steam engines here, for the mill was already set up for water power. Two splendid breast shot wheels, each 15ft 6in in diameter and 11ft 6in wide were now used to work a pair of beam pumps. It is wonderfully ingenious, to use the Avon itself to lift the water nearly 50ft up the hillside to supply the canal. Claverton pumping station is preserved, though a modern motor generally takes care of the work. In 1989, however, modern technology failed and the old system was brought back into use and proved it could still do the job. In its working days, Claverton water pumps were shifting 100,000 gallons of water an hour.

The next important point along the way arrives at Dundas. Here the natural lie of the land forced the canal across the Avon, and Rennie designed an aqueduct that can be thought of as his homage to the architecture of Bath. He used the lovely, golden Bath stone, and crossed the river in a single arch, with relieving arches to either side. Pilasters ornament the spaces between the arches and the whole structure is topped by a deep, dentilled cornice and a balustrade. It is the aqueduct that catches the eye, but Dundas was an important junction in its day as well. In 1794, work began on a canal to join the Kennet and Avon to the Somerset coalfield. The entrance lock and lock cottage can still be seen, but the canal itself now runs only a short way to a hire boat base. The Coal Canal itself will be discussed later (p.86). Some idea of the importance of the junction can be gained from the extensive wharf with its hand crane.

The next section is a delight to modern pleasure boaters, but was a nightmare for engineers. The clay puddle was constantly being undermined by water from the hillside, and the section was often drained for repairs, hence the large number of places where the canal narrows down for stop planks to be inserted. The excursion to the north bank of the Avon is a short one, and soon the canal is crossing back again at Avoncliff. This is a notably less elegant aqueduct than Dundas, with low arches and the local ragstone has not worn well. The stone itself came from quarries near Westwood, which were linked to the canal at this point by tramway. There is also a first indication that this rural region was once an important industrial area. Water-powered mills can be seen at each side of the river: one, with a distinctive square brick chimney is a former woollen mill and the one on the opposite bank was a flock mill. The canal is, in fact, nearing Bradford-on-Avon and the end of the long pound. Bradford was in its day a very prosperous wool town. Visitors come to admire its picturesque charms, but are not always aware that the grandest houses belonged to clothiers, and a little wander down the river will reveal a whole series of mills. The canal itself passes well above the town, but recognises its importance with warehouse and dry dock. Now the canal begins what will prove to be a long climb to the summit.

The canal follows a very quiet route through the Wiltshire countryside, with a background of gently swelling downland. At Semington an insignificant opening marks the start of the

The Kennet and Avon Canal at its most debonair as it passes through fashionable Sydney Gardens in Bath.

Wilts and Berks Canal, begun in 1795, and heading off for Swindon and the Thames at Abingdon. It was abandoned in 1914, but that has not stopped enthusiasts drawing up plans for its eventual restoration. Everything is deceptively peaceful, and nowhere more so than at Seend Cleeve, as the canal weaves and climbs through a landscape of fields and wooded knolls. Yet a century ago this was one of the busiest places along the route, home to Seend ironworks, where the blast furnaces were turning out 300 tons of cast iron a week. Soon, however, the rural idyll that boaters have been enjoying comes to a brutal end. The canal charges up the hillside to reach Devizes in the mighty Caen Hill flight of twenty-nine broad locks. The most impressive section lies between locks 29 and 44, where they follow each other in rapid succession in a straight row – a truly daunting sight. To reduce water loss on the flight, each of these locks has a large side pond, terraced into the hillside, so that it really does look like an immense watery staircase. The effort was justified by the importance of Devizes to the canal. An important source of cargo was the large brickworks near the flight There were three wharves in the town. Town Wharf retains one of its old warehouses, since converted into a theatre, while Devizes Wharf has an imposing range of buildings that have in turn served as granary, bonded warehouse and, now, the Kennet and Avon Canal Trust offices and exhibition centre.

Once Devizes has been left behind, it is very much a case of out into the country again. Settlements along the way are either quite small, like delectably named Honeystreet, with a typical small grouping of wharf, warehouse and pub, or missed altogether. Pewsey was not thought worth a diversion, so another small settlement was established half a mile away at Pewsey Wharf. All this means that once again the engineer could keep his canal on the level – which after Devizes is certainly welcome news to boaters. Fifteen miles go by before

Wootton Rivers is reached and three widely spaced locks lead up to the summit. Bruce Tunnel is something of an absurdity, since it does no more than dip beneath the surface, and it would have been perfectly possible to continue the deep approach cutting right through the low hill. The answer, as is so often in these cases, was the need to appease a powerful landowner. Thomas Bruce, Earl of Ailesbury, did not want to see the canal as it passed through his land, so it was cheaper to build a tunnel than go through the expense and time-wasting negotiations that would have been needed to achieve engineering logic. In the long term, it might have been better to hold out, as the tunnel has no towpath and proved a bottleneck.

This is a short summit, so once again the question of water supply had to be tackled. Nearby Wilton Water was to be the source, and in order to lift the water up to the canal, a pumping station was built at Crofton. Here two beam engines were installed, the first having been supplied by Boulton and Watt in 1812. This is the oldest steam engine in the world, still to be found on its original site, quite capable of doing the job it was installed to do nearly two centuries ago. The engines have been fully restored to working order, and they are among the glories of the steam preservation world. Right at the top of the house, the cast iron beams nod in stately motion. One floor down are the cylinders and a gleaming array of valve gear, while down at the lowest level is the condenser, hissing and spluttering like a witch's cauldron. Steam is still raised in coal-fired boilers. The power of the engines is evident as the water pours out into the leat for its journey to the canal. Working together these two giants could shift a million gallons of water an hour.

The canal now makes its way through open country to Hungerford and the Kennet valley, where the going gets easier as the canal closely follows the line of the river. There are reminders of Bath in the approach to Hungerford as the canal passes the pinnacled tower of

Different engineers, but similar lines: Rennie's Kennet & Avon takes a higher level route above the Avon than Brunel's Great Western Railway.

A peaceful scene at Seend: it is difficult to imagine that this was once the site of a big iron works.

Hungerford church. The old church collapsed under snow in 1812, and the Bath stone for its replacement was shipped in by canal. This contrasts with the other buildings, for this is a clay area and the older houses are often timber framed and the canal structures are brick. There is one more big town, Newbury which was, like Bradford, a busy centre of the woollen trade when the canal was new. The canal ends at Newbury lock with its attractive lock cottage, as it joins the Kennet Navigation.

Turning back now to Dundas and the junction with the Somerset Coal Canal brings a waterway which is derelict and likely to remain so. That does not mean, however, that it is devoid of interest. Much of the towpath can still be walked as a public footpath, and provides a view of some interesting structures – and an intriguing puzzle. The most rewarding section begins at Midford, half a dozen miles from Dundas, where the canal can be joined near the Hope and Anchor pub. Just beyond this point, the canal divided, one branch heading off to Radstock, the other to Paulton. The towpath follows the latter, but just to the south is the

The hugely impressive Caen Hill flight of broad locks with their array of side ponds at Devizes.

Crofton pumping station. The engine house with its tall chimney contains a pair of steam engines that pump water into the canal from the lake, Wilton Water.

aqueduct that brought the Radstock branch across the Cam Brook. The reasons for not following this particular line are tied in with the history of the waterway. Work in the early years had extended the canal from Radstock to Wellow, while at the Midford end, basins and wharves had been created not far from the end of the aqueduct. The steep hill in between was thought to require eighteen locks in about one mile of canal, and as a 'temporary' measure a tramway was constructed instead. It was to become a permanent fixture, and in 1814 the decision was taken to build a tramway all the way from the Midford basins to Radstock and the collieries. So, not surprisingly, the Radstock canal was soon only a memory. The Paulton line made its way up the valley of the Cam Brook towards the foot of the hill at Combe Hay, through a series of locks, which can still be seen as empty chambers. Then it goes through a hairpin bend at what was known as The Bull's Nose. This often rather swampy corner beneath the wooded hill, known as Engine Wood, was once home to a steam pumping engine, and is a section of canal that has seen immense changes. Somehow or other the canal had to get to the top of the hill. A solution was offered by an ingenious engineer called Robert Weldon: the caisson lock. The caisson was a watertight box, fitted with doors at either end. It floated in what was, in effect, a large lock, though it was more like an oversized well. The caisson could rise and fall in the lock, by pumping air in or out, its motion controlled by guide rods. Boats entered the lower level by a short tunnel, floated into the caisson, were closed in and rose to the surface to be released onto the upper level. The process could, of course, be reversed for going downhill. It was ingenious but impractical, and was soon abandoned. The upper level stood near a big house, still called Caisson House and excavations have identified the site, but sadly not enough emerged to give a more detailed idea of how it was actually constructed. With the demise of the caisson lock, the next step was to build an inclined plane, which can still be seen as a track through the trees and can be followed up to Combe Hay. By 1802, the incline too was to be replaced by conventional locks – though even these had their interesting features, for one can still find traces of iron lock gates. For anyone with a well developed detective instinct, this is as good a walk along an old canal as you can hope to find. The remainder of the line was altogether more conventional. It is easy to look at a derelict canal and think that it must have been a rather insignificant affair to deserve closure. This is certainly not the case with the Somerset Coal Canal, which provided a good deal of the traffic for the western end of the Kennet & Avon, and only fell into disrepair because much of that traffic was taken away by the railways. The other canal built in southern England in the 1790s may look much grander, a fine wide barge canal, but in fact the Basingstoke was never going to be as prosperous as the modest Coal Canal.

The first Act for the Basingstoke Canal was passed in 1778, but little progress was made and a second Act was passed in 1793, allowing the company to raise the funds to finish the work. There is an interesting sidelight to the work of construction. The contractor was John Pinkerton, and instead of paying his men in coins of the realm, he issued them with tokens which could be exchanged for goods and, in theory, for cash. The Basingstoke Canal token has the name of the canal on one side, with a picture of a square rigged sailing barge, and on the other Pinkerton's name and the navvy tools – barrow, shovel and pickaxe. The canal was to run from Basingstoke itself to a junction with the River Wey at Woodham, a distance of 37½ miles with twenty-nine locks. From the start, it was never altogether clear where the traffic was to come from to keep this largely rural canal in funds. Ironically, one of the canal's busiest periods came in the 1830s carrying material for the construction of the London & South Western Railway. The one steady source of cargo came rather too late in the day, in the 1890s, when brick works were opened at Up Nately. Nothing, however, could prevent the steady decline in trade, and with no funds for repairs the waterway deteriorated to the point where it was virtually abandoned. Yet it had its admirers, and ranks among the first to be considered for restoration. One of the standard reference books used by canal historians is the canal Bradshaw, Henry de Salis' *Handbook of Inland Navigation*. My own copy is a 1928 edition, and on the pages giving the details on the Basingstoke, someone has added an amendment in ink. 'Bought on behalf of Inland Waterways Association first week in March 1949. Objectives stated by buyers to preserve amenities for Ramblers, Hikers and for Pleasure Craft, as well as for Trade', and then gives the name of the lady who purchased the canal on their behalf, Mrs Marshall of Fleet. It was a bold venture, but led nowhere. In time the local county councils stepped in to purchase the canal, and volunteers set to work in earnest. Those who took part in the great Deepcut Dig will all have their own memories of battling against a sea of black silt. My own recollection is of wheeling barrow loads of the stuff along planks that became increasingly indistinguishable from the surrounding goo. At one point the barrow missed the plank altogether – or, to be fair, its handler should take the blame. It proceeded to sink out of sight and the handles were just grabbed back in time before it disappeared altogether. At least the work was done.

The start of the canal at the Wey is now overshadowed by the M25, and it struggles to get clear of the suburbs. The canal climbs steadily at first, then comes the group of five St John's locks, followed by three Brookwood locks and a flourish of fourteen at Deepcut. This was where the heavy engineering work came in. The canal did not arrive at a place called Deepcut – the place took its name from the work. It is an impressive place, a cutting stretching for over half a mile and in places some 70ft below the surrounding land. It has a somewhat forbidding air, a dark gash in the land, overhung by beech and chestnut, so that it seems to be permanently dusk. At this point the railway comes very close to the canal, and in these early days of railway construction, the problem of 'frightening the horses' was a very real concern. So wherever the line came within a 100ft of the towpath, the railway company had to build a high brick wall to prevent any possible accidents.

The rest of the line was built almost on the level, with just one solitary lock at Ash in twenty-seven miles. This involved some serious diversions. Beyond Deepcut, the canal arrived at the Blackwater valley and rather than cross it immediately, which would have either involved immense earthworks or a lot of locks, the canal headed south along the rim of the valley until a point was reached where a crossing seemed easier. Spoil from Deepcut was boated down, and an embankment built across the canal at Ash. It was to become one of the victims of neglect, crumbling away, creating a breach that drained much of the canal. It has been replaced by an aqueduct. The engineers of the L & SWR had no worries about extensive earthworks. They were quite happy to create a deep cutting that would emerge near the valley floor. They did have one problem, however – the canal. It was higher up the hill, so they decided to build a new aqueduct to carry the canal over the railway cutting. It was originally carried on two arches, but when the track was increased to four lines, the length had to be doubled and two extra arches added. To avoid closing the canal during construction, they built the aqueduct to double width, so that one side could remain open while work went on with the other. A canal aqueduct constructed by a railway company is certainly a rarity.

Having crossed the river, the canal engineers could set their sights on the end of the line at Basingstoke, more or less due east, but before getting there, there were to be more convolutions as the canal hugs the contours in a way scarcely seen since the days of Brindley. At

Lock and lock cottage on the Woodham flight of the Basingstoke Canal. This very rural canal was built with broad locks to take barge traffic.

Even the names on the Basingstoke seem to underline its rustic nature: this is Barley Mow bridge at Winchfield Hurst.

first the route heads more or less straight towards its objective, through what is now a very militarised area. To the south is Aldershot, famous for its army camp, which during its construction in the 1850s provided the canal with a good deal of trade, bringing in brick and timber. To the north is the later, and equally famous, Farnborough airfield. The canal leaves the urban surroundings for an area of open heath and woodland before a short run through Fleet brings it at last to wholly rural surroundings. This is a landscape of humps and hollows, many of the latter having filled to create ponds and flashes – which were to prove very useful sources of water for the long summit level. It is all very pleasant but inconveniently lumpy, and the canal goes round everything with the minimum of earthworks. A road runs from the canal at the outskirts of Crookham and rejoins the waterway just beyond Dogmersfield, a distance of a mile – boaters have to travel nearly three miles. At least this is an area where it is a pleasure to linger and the villages that do crop up along the way are places that generally offer a welcoming face. At Odiham, for example, as so often with contour cutting, the canal

The beautiful scenery at Deepcut locks, the scene of much hard labouring during the early years of restoration.

Brick Kiln bridge stands at the end of navigation on the Basingstoke.

builders were unable to get near the centre, so the town had to come out to meet the canal. Here is the typical scene of a bridge of mellow brick, very much the standard design, and canalside pub. Everything holds together, the use of local materials providing the unifying theme. Now a new marina offers day boat hire and a boat trip, and a second bridge over the canal, carrying the bypass. Comparing the old and the new is an interesting exercise.

The canal engineers were about to reach the greatest obstacle to progress. Ideally, they would have liked to swing north round Newnham to take the easier ground, but Lord Tynley who had the manor and most of the land, would have none of it. To the south the way was barred by a ridge at Greywell, and now there was no alternative – too long to go round, too deep for a cutting – there would have to be a tunnel. It was 1,230 yards long, and there was at least one apparent advantage gained: during the construction springs were found in the chalk, which added to the water supply from the ponds along the way. Unfortunately, springs and stability rarely go together. In 1932 part of the tunnel collapsed, and the whole of the western end of the canal was abandoned. The Basingstoke Canal no longer went to Basingstoke. The whole section of the canal was sold off, the old town basin filled in and a bus station built on the site. Today the canal ends by the abandoned lock 30, built after the canal was completed to raise the level to improve navigation through the tunnel. The tunnel may no longer be busy with boats but it has a huge population – of bats, and has been designated as a world wildlife heritage site. The end of navigation does at least have two engineering features, the modern lift bridge at North Warnborough and the modest aqueduct over the Whitewater, and it does enjoy a bit of history. Overlooking it is the octagonal keep of Odiham Castle, said to be King John's last stopping place before setting off for Runnymede to sign the Magna Carta.

7. Manchester and the North

A great deal had happened since the Duke of Bridgewater built his first canal into Manchester. The world had begun to go through an industrial revolution, with cotton at the heart of it. From beginnings in Derbyshire, the new textile mills had spread across to Lancashire, and Manchester had developed from a modest town to a very large town, and would eventually become a city. In the 1760s it had been modest. By 1784 a visitor was describing it as a 'large and superb town', and one that had been largely built over the past twenty years. It was big in 1784 – and it would be twice as big by the end of the century, and would still be growing. Such a huge rate of development brought new demands for goods: coal for fires, building materials for development and, of course, raw cotton for the mills. It was inevitable that Manchester should be the meeting point for a new network of canals.

Among the first of this new generation was the Ashton Canal, which was originally designed as a simple link between Ashton-under-Lyne and Manchester, but which was destined to spawn an array of branches, and make connections with three other canals – and all this in a main line, less than seven miles long. This was a canal that was immensely successful in its day, with numerous short arms to commercial wharves at collieries and factories. There were branches to Islington, Stockport, Hollinwood and Fairbottom. They thrived when canal traffic thrived, but went into decline together, so that by the 1960s they were all faced with closure. In fact, the protest by waterway enthusiasts, who sent a flotilla down the canal in 1961, proved that it was closed in fact, if not in official documents. A lock gate had been lifted off and thrown into the canal by vandals. It was not until 1974 that a boat again made a trip down the whole canal, and that year it was officially reopened – though minus its branches. One can see why the authorities wanted to see the canal closed: it was never going to be a candidate for a beauty prize. In the years since it was completed, Manchester has spread to the point that it is virtually impossible to determine where it ends and Ashton begins. Industries have crowded in on the waterway, but the collapse of the cotton industry saw what had once been prosperous manufacturing areas degenerate into urban wastelands. It was not, it was argued, a canal that anyone would want to visit for pleasure. What the would-be closers had not appreciated was the vital role played by the Ashton as a link between other waterways. It was this that justified the expense and effort of restoration.

Inevitably, much that gave the canal its original character has been swept away. The start is at Ducie Street, near the centre of Manchester. Once this was a busy place with basin and warehousing, but the once imposing warehouse complex was destroyed by fire. There are still canal offices, but these belong to the Rochdale Canal, which arrived to make the terminus into a junction. There is a short aqueduct before the canal begins its climb through the three Ancoats locks, hemmed in by the often drab and derelict factories. There is no point in complaining about the surroundings: this was why the canal was built – to serve industry. These are the monuments to the canal's success in the nineteenth century, when it acted as the 'motorway corridors' do today, offering prime sites with first class transport facilities. During the restoration, every effort was made to make this a showpiece of how canals could be made attractive features of an urban scene, though the planners did not always seem to understand how locks were used in practice. I still remember following the usual canal practice of opening a lock by putting my back against the beam and walking backwards, only to fall over a bollard set directly in the way. There were twenty-five locks in all, eighteen on the main line ending at Fairfield Junction, and the rest on the 4½ mile long Hollinwood Branch, now closed. Fairfield Junction provides an interesting spot, a place to study contrasts in materials and styles. At the tail of the last lock is a long, low arched bridge, built of stone with a cobbled pathway on top. This was the material most readily available for builders in the canal age in this area. But the spread of the waterways and later the railways, made it possible to use cheaper materials. So you find local factories, the steam powered cotton mills, built of an often garish red brick. They can seem uncompromisingly ugly, but they were not built with aesthetics in mind. It is easy to point to the graceful elegance of the canal bridge and praise it as altogether more attractive. It may be more beautiful, but its builders were no more interested in its visual appearance than were the mill architects. It uses curves because the arch is traditional for bridge building, is economical in materials and the gentle curves make life easier for the canal horses. The mill builders used simple, block-like structures to provide the maximum space for machinery.

Fairfield was the original terminus of the Ashton canal, and a branch line was constructed to extend the Ashton to meet two new arrivals on the canal scene, at two further junctions. Until recently, just one of the connections led to a still navigable canal, the Peak Forest. In 2001, however, the second connection was made with the reopening of the Huddersfield Narrow. At the time of the restoration, however, it was the fact that the Ashton connected with the Peak Forest, and that together they could form part of a cruising route which came to be known as the Cheshire Ring that gave the impetus for renewal. So the next canal to look at is the Peak Forest.

The canal was seen from the start as feeding into the Ashton, and was itself to be nurtured through tramways, or as the original Act has it 'Railways or Stone Roads'. There has been some confusion about the name, as there is a village called Peak Forest, but the canal not only never went there but was never even meant to go there. In fact, Peak Forest was then not a single place, but a wide area. The start is at Dukinfield Junction, closely followed by a short aqueduct over the River Tame. At first the Peak Forest is not as closely surrounded as the Ashton by mills and warehouses, but the gains of having an open vista are slightly reduced by the fact that most of the space is used for rubbish dumps and sewage works. After that, there is a succession of small towns, each boasting its cubic brick mills, some with names picked out in liverish, yellow brick. The canal has so far been keeping to the valley of the Tame, but as that river turns away to the west, the canal has to find a way through more difficult ground. There are still no locks en route, but two tunnels had to be constructed. The first is the comparatively modest Woodley Tunnel, 167 yards long and enjoying the luxury of a towpath. The second at Hyde Bank is quite different. It is unusual in having an elliptical arch at the entrance, is lower than Woodley and has no towpath. The discrepancy can easily be explained: as with so many other canals the Peak Forest, begun in high optimism in the mania years, ran out of cash. A towpath became an unaffordable luxury. It is possible that cash shortages resulted in skimping in many ways, for the tunnel is not well constructed, being very far from keeping to the straight and narrow. There was a third short tunnel, but this was opened out to create a cutting, leading on to the start of a whole series of spectacular engineering features.

Ancoats on the Ashton Canal: the lock refurbishment has followed traditional lines, with setts to provide a good grip when moving the balance beam.

The canal now heads towards the next river valley, not to join it, but to cross it. The Goyt lies in a deep valley, and the engineer Benjamin Outram, working with Thomas Brown, designed a splendid aqueduct over the river. There are three arches, the central rising 100ft above the Goyt. In an area of abundant stone, there was never much question of what material was to be used. Although Outram was himself a partner in the successful Butterley Iron Works, he made no attempt to use this material, not even for the trough lining, which was made watertight in the traditional manner using puddled clay. There is an immense amount of material in the structure, but some lightening of the load was achieved by using another new technique, pioneered in the 1760s by William Edwards of Pontypridd. He built a bridge that crossed the River Taff in a single 140ft span arch, but had great difficulty in completing the work. The arch collapsed twice, and on his third attempt he relieved the pressure on the centre of the arch by piercing the abutments with cylindrical holes. This solved the problem, and the device is used here: the effect is decorative, but wholly practical.

Once across the aqueduct, the canal engineers were faced with a steep climb to the town of Marple. At first, there were no funds to pay for an extensive flight of locks, so a plateway was built instead. It was only after a second Act was passed in 1800 that enough money could be raised to complete the work, so that the plateway continued in use right through to 1804. It certainly is a daunting prospect, for the canal had to be raised by 210ft, and this was achieved in a flight of sixteen locks. The locks are packed close together, and there is a chance to compare the canal solution to conquering this hostile landscape with that of the later railway engineers. The line from Marple Rose Hill Station arrives on a high embankment which ends at a lofty mutli-arched viaduct. Everything along the canal has a robust quality, constructed of roughly-dressed stone blocks. One of the most interesting buildings is the warehouse built for the local cotton magnate, Samuel Oldknow, who was a major investor in both the canal and the connecting turnpike roads. It was built to make the best use of both canal and road connections. At the landward side, there is the normal arrangement of loading bays and hoists, but boats could come in under an arched entrance to be loaded and unloaded under cover. There is a small toll house by the warehouse, where road users paid their fees. It is all very well planned for economy of use, and nothing has been spent on architectural niceties, but then it has no need of them. The patterns created in stone, used throughout for walls, lintels and mullioned windows are visually satisfying without any embellishment.

At Marple, the locks are crossed by a busy main road. So, to avoid problems, horses were led through the bridge abutment in a special horseshoe shaped tunnel. The canal is now joined by the later Macclesfield, one of the last to be built before the world turned away from canals, and railway fever took over from canal mania. There are indications now of just why this canal was built in the first place. Limekilns occupy a prominent canalside site, and these were in fact also part of the Oldknow industrial empire. The original intention, as set out in the Act, had been to take the canal to the limestone quarries at Chapel Milton, near Chapel-en-le-Frith, on the edge of what is now the Derbyshire Peak District National Park. The climb up to the quarries proved a step too far for a waterway, as we shall soon see. As it was, the canal had to be taken on a high level route, clinging onto the steep hillside, in a way reminiscent of the Brecon & Abergaveny. It shared with that canal a tendency to landslip, and there were serious breaches in the 1940s and again in the 1970s. Shortage of funds is reflected in the swing bridges along the way, cheap in the short term – more expensive and troublesome in the long. It is some years since I last boated along this section, but even now I can remember the hard labour involved in shifting some of these obstinate bridges.

The plans to extend the canal up the valley were abandoned. Instead a plateway was constructed from Buxworth – which was then known by the less genteel name of Bugsworth – for 6½ miles. This was the original terminus of the canal, but a second was created following the construction of the Cromford & High Peak Railway, linking this canal to Cromford (See Vol. 1). The railway and canal met at an interesting interchange at Whaley Bridge. The terminal warehouse is a simple structure, built over the end of the canal arm, so that boats could float inside. What makes it unusual is the opening at the far end, from which railway lines can still be traced, heading up over the Derbyshire hills. The original terminus at Buxworth became derelict after the tramway was closed down in the 1920s, but has recently been largely restored by volunteers. It is a magnificent complex. The positions of railway sidings are marked by the lines of stone sleeper blocks, running alongside the basins and wharves and out to the old lime kilns. The canal structures are themselves impressive, particularly a stone bridge, with a cobbled path running across the top. It is not hard to imagine what the site must have been like at the end of the nineteenth century, when as many as forty boats a day were leaving Buxworth with loads of lime or stone. For those who would like a little help in depicting the scene, the local pub on the wharf has photographs of the working days.

After this excursion into Derbyshire, it is time to turn back to Manchester and two very remarkable canals, built across the Pennines. They both received their enabling Acts on the same day, 4 April 1794, but one was to open seven years before the other. The 'speedy' winner was the Rochdale, which was completed in 1804: the slowcoach was the Huddersfield Narrow, which finally opened for business in 1811. The earlier trans-Pennine route, the Leeds & Liverpool, had largely avoided the inherent problems of the hills by opting for a wandering route that took it far to the north; the newcomers were bolder. The Rochdale begins at Castlefield Junction, linking it to the Bridgewater and proceeds via a second junction with the Ashton to reach the Calder & Hebble Navigation at Sowerby Bridge. The thirty-three mile journey involves the negotiation of ninety-two locks. The Huddersfield Narrow also joins the Ashton, at Ashton-under-Lyne, and its journey to reach the Huddersfield Broad Canal in Huddersfield is quite short, just twenty miles. But in that distance there are seventy-four locks and, crucially, a tunnel under Standedge Fell. It is three miles 135 yards long, the longest in Britain – and it has no towpath. Astonishingly, according to Priestley's canal guide of 1831, the leggers completed the journey in 1 hour 20 minutes. It was the building of this long, deep tunnel that created the delays, and ate up the company's funds at an alarming rate.

The start of the Rochdale at Castlefield Junction was, until quite recently, a rather forlorn, run-down sort of spot, but not without its unexpected pleasures. I once clambered up onto the old railway line above the basin to find the track blazing with a mass of wild orchids. Now it has all changed and become the heart of new, trendy Manchester. But moving out from the centre, the canal resumes its old character, isolated from the city streets, lost from view behind high stone walls. At once the climb begins, and in this cramped urban setting there is not even room for such amenities as overspill weirs. The result is that in very wet weather, the water has nowhere to go, and locks are all but submerged. The modern world does not merely encroach on the canal, in one place at least it overwhelms it. A modern tower block straddles the waterway, and there in a thicket of concrete trunks is a lock, surely the gloomiest, darkest lock in Britain. On every hand there are memories of the days when Manchester was the cotton capital of the world: in street names such as Bengal Street and

The magnificent Marple aqueduct carries the Peak Forest Canal over the River Goyt. The holes pierced through the spandrels of the arches lighten the load without weakening the structure.

An exceptionally fine stone bridge at Buxworth Basin, an important centre on the Peak Forest Canal and its original terminus.

China Street, telling of international trade; in often ornate warehouses, built in a bewildering variety of styles from Victorian Gothic to Indian Moghul. This section of the canal has been open for a long time, a vital link in the Cheshire Ring, but it comes to an end at Ducie Street Junction. This was once presided over by immense, brooding warehouses now demolished, but it is still home to the canal company offices. At the time of writing, major restoration work is still in progress, with two very big obstacles to be overcome: the site where the canal was filled in and a supermarket built on top at Failsworth, and the total blockage of the canal by the high embankment carrying the M62. The new works needed to complete restoration will be discussed in the next volume, looking specifically at later canals and new works. The description that follows will be looking at the original works on the Rochdale.

The engineer responsible for the canal was William Jessop, and in many ways this was his most challenging task. It is one thing to climb over the Pennines with locks, quite another to find sufficient water to feed them, given that the summit level is less than a mile long. An added difficulty lay in the fact that this was no narrow canal, but one designed to take vessels of ample dimensions, 14ft 2in beam and 74ft long. The first plans for the canal, drawn up by John Rennie, had called for a summit tunnel, 3,000 yards long. Jessop was not enthusiastic – and given the troubles created by tunnelling on the Huddersfield Canal, his scepticism was justified. So the summit was set in a deep cutting, though an extra seven locks were still needed to take the canal a little nearer to the hilltop. All this demanded a huge amount of water, and available natural sources were limited by the concerns of textile mill owners along the route, who feared for their own water supplies. It was agreed that their supplies would be protected and none of the water for mill feeders would be diverted to the canal. It was

The Anatomy of Canals

Dukinfield Junction, where the Ashton and Peak Forest meet is graced by a bridge which carries the towpath of one over the other in a sweeping curve.

obvious, in these circumstances, that reservoirs would have to be built, and Jessop originally planned for two, at Hollingsworth and Blackstone Edge, while a third was later added in the Chelburn Valley. It is something of a curiosity that reservoirs such as these, which must have been scars on the landscape during construction, and for many years afterwards, are now thought of as beauty spots. Hollingsworth is now Hollingsworth Lake and centre of a country park. What is not obvious is the huge amount of work that went into its creation. The bank of the reservoir was built 10ft wide at the top to take a roadway, and it was constructed on a 1:2 ratio. To keep it watertight, a 9ft thick core of puddled clay was included. Even with such massive works, it was still considered prudent to minimise water loss. Jessop designed the locks, eastward from Ducie Street, to have a uniform rise and fall of approximately 10ft, so that, in his own words, 'the Water which serves one, will serve the Rest without Waste'. This is not quite the end of the water supply story. There are many places along the canal where hill streams could be tapped for extra supplies. These streams inevitably carry sediment, which if allowed to fall into the canal would result in increased dredging costs. So Jessop designed settling tanks at the canal side, which could be easily cleared before the stream actually reached the waterway. The Rochdale Canal remains a superb example of Jessop's ability to combine a good overall design with care for every detail.

The same care for detail can be seen in the excellent stonework of the many bridges along the way, not least in the two skew bridges, Gorrell's and March Barn, between the Rochdale and Heywood Branches. Most bridges are designed to cross a canal at right angles, and where older canals met roads, either waterway or road had to be bent to make this possible. The alternative, to build a bridge at an angle, is more complex. At Gorrell's Bridge, this is achieved by using massive, overlapping stone blocks, but March Barn uses the far more sophisticated technique of laying the stones in winding courses. Seen from underneath, the stones appear set in a series of diagonals. Jessop was the pioneer for this very useful technique, which was to be used on many later canal and railway bridges.

The climb up from Manchester is very much dominated still by the old cotton mills, many built in the violent red brick, known as Accrington brick, very popular with nineteenth century and early twentieth century builders. Stone, however, still dominates the canal structures and the older buildings. Lock 37 marks the end of the climb to the summit. The railway which has been keeping the canal company for some way dives into a tunnel, emblazoned with the arms of the Lancashire & Yorkshire Railway. The railway engineers did not enjoy the luxury of climbing the hill in a series of steps, and there was a limit to the gradients that could be overcome by nineteenth century steam locomotives. The summit itself is a wild place, where rocks poke through the crust of the moor, and a sturdy lock cottage stands beside lock 36.

Up to this point, the canal has been generally heading north to find a comfortable passage through the hills, but now as it begins to descend, a steady swing to the east begins at Todmorden. There is now an obvious route to follow along Calderdale. The railway has re-emerged and crosses the canal on a conventional bridge, enlivened by castellated stone turrets at both ends. The moorland setting, the succession of locks and the imposing bridge make for a picturesque scene, where even the man-made structures seem at home, thanks to the use of local materials. Todmorden itself provides a fine chance to enjoy the older style of mill town, all very Lowryesque, with cobbled lanes, mills and warehouses. The canal passes through the middle so that the boater gets to see it all. The next section, through to Hebden Bridge, boasts some of the best scenery on the canal – and posed some of the greatest problems for the engineer. The river runs in a narrow cleft through the hills, with the south

The summit of the Rochdale Canal is reached after a climb of thirty-seven locks from the east and fifty-six from the Bridgewater Canal in the west.

dominated by a craggy hillside along which the canal had somehow to be built. A way had to be blasted through the rock and a shelf created into which the canal itself could be sunk. There is an interesting pub at Stubbings Wharf, with direct access from the towpath at first-floor level. It is, as far as I know, the only canalside pub to be celebrated in verse. The poem *Stubbing Wharfe* by Ted Hughes describes a dour pub full of dour people, where he and Sylvia Plath singularly failed to find any sense of pleasure at all: happily, it has become altogether cheerier these days. Though I am still wondering how a Norfolk wherry came to be included among the local canal photos on the wall! This stretch of waterway also provides a rare chance to see a horse-drawn narrow boat, carrying passengers rather than cargo. It is instructive to see how a horse boat negotiates the locks, without the valuable aid of a reversible motor for stopping. Hebden Bridge is approached via a short aqueduct over a tributary of the Calder and has an extensive, redeveloped wharf. This is an intriguing town, built up a steep hillside, where houses are top to bottom, rather than back to back. Seen from below the houses seem to be terraces of four storey buildings, but appear as conventional double storeys when seen from above. This is a town which grew with the canal age, gradually usurping the older woollen centre of Heptonstall on the hilltop. The canal itself played an important role in the change, not least because it acted as a drain in what had been a very swampy valley.

Beyond Hebden Bridge, the valley begins to widen out and the going gets easier, though there is still the same relentless march of locks. The route ends at Sowerby Bridge, with a large basin and trans-shipment warehouses, where cargoes could be exchanged between the short barges of the Calder & Hebble Navigation and the wide boats of the canal. This is now a hire base, but the old warehouses with their boat holes remain as impressive reminders of the working past.

Just beyond the lock at Hebden Bridge is the aqueduct carrying the Rochdale Canal over Hebden Water, and just a few of the cotton mills that line the canal.

Having reached Yorkshire we can now travel back again on the Huddersfield Canal. The canal was begun at the same time as the Rochdale, but it was not completed until 1811, suffering from a mixture of financial and engineering problems. This is a narrow canal, taking the shortest route of the three trans-Pennine waterways, but the price to be paid for the direct route appeared in the form of Britain's longest canal tunnel, all 5,415 yards of it. Even allowing for this, it still needed a total of seventy-four locks in less than twenty miles to bring the canal to the highest summit in Britain, 683 ft above sea level. But, if it lagged behind the Rochdale in opening in the nineteenth century, it won the race for re-opening, when restoration was completed in 2001.

The canal begins at a junction with the Huddersfield Broad Canal (Vol. 1) in Huddersfield. The route taken is an obvious one, following the line of the Colne Valley through a succession of mill towns. This is the line taken by the main road and the later railway, all crowding into the narrow valley bottom. Just as the route makes sense in engineering terms, never straying far from the river, so too it appeared to make financial sense, for there was no shortage of customers along the way. This is an area with a long history of woollen manufacture: the older houses with their long weavers windows recall the pre-factory age when hundreds of handlooms were in use in the area. They can be seen in towns and high on the moor. The later mills stay close to the valley floor, crowded together at the Huddersfield end, but spread out all along the line. There is a particularly interesting group between Slaithwaite and Marsden, where the canal threads a path between reservoir and mill pond. The mill itself is isolated, with a small terrace of cottages for the work force. As the canal makes its way west, the hills crowd in on the sides, their tops marked by lines of crags and quarries, supplying the millstone grit which gives the area its characteristic look, the dark stone of mills and houses echoing the dark lines ruled along the hilltops. Beyond Marsden, the valley ends at the brooding mass of Standedge Fell, rising to a height of almost 1,500ft. There was no way over that for a canal, so it had to be forced through. The approach is via a cutting made deeper by the spoil banked up from excavations first for the canal tunnel, then for subsequent railway links alongside. The maintenance yard is now a canal museum, and just beyond that is the tunnel keeper's cottage and the modest opening in the hillside, that marks the start of Standedge Tunnel.

The tunnel was the greatest challenge facing both the original builders and the later restorers. The initial construction problems were so great that the decision was taken that it would be built as a single bore with no towpath, just as Brindley's pioneering Harecastle Tunnel had been built on the Trent & Mersey many years before. Anyone exploring the canal on foot can still follow the path taken by boat families in the working days, when they led their horses over the fell. It is possible to miss this out by driving to the far side, or even by taking the train, but there is actually a lot to see on the way – as well as being a most enjoyable walk. The tunnel was constructed by driving headings from the bottom of a number of shafts,

The Huddersfield Narrow Canal heading down the Colne Valley from Marsden to Huddersfield – a descent that will require forty-two locks in all.

The restoration of the Huddersfield Canal required a whole new section to be constructed through Slaithwaite.

many of which were retained for ventilation. There are also remains of the engine houses for the four steam engines that were needed to pump water out of the workings. The biggest, which rears up above mountains of spoil, can be seen near the Redbrook Reservoir, which was also built by the canal company. A trip boat provides an opportunity to see part of the tunnel, but those wanting to take their own boats right through have to wait for the electric tug. As the long journey gets under way, it is worth thinking what it was like for those who came this way before the motor, when the boats had to be legged through. That was not the only problem: a new hazard appeared with the coming of the railway age. Some years ago, Charles Hadfield admitted that he was not in favour of reopening the Huddersfield, simply because of his memories of boating through in the days of steam on the railway. Railway and canal shared ventilation shafts, and each passing train produced clouds of foul, black smoke which billowed out into the canal in choking clouds.

The tunnel ends at Diggle, and there is a curiosity in that it bears the date 1893. The canal was in fact realigned and a new section created by the method of 'cut and cover' – digging a deep cutting and roofing it over. The most dramatic point on the western end of the canal occurs on the outskirts of Uppermill, where the canal crosses the River Tame on a low-arched aqueduct, while they in turn are all but overwhelmed by the splendid railway viaduct. The technique of building a skew arch was well learned by the time the railway engineers came along, and there is a very fine example high above the canal. There is one further tunnel along the line, Scout Tunnel, which is short and has the benefit of a towpath, even though it is still not wide enough to allow boats to pass. The locks still come in steady progression, thirty-two of them dropping the canal 338ft, quite modest compared with the forty-two locks raising the waterway 439ft on the Yorkshire side. Inevitably, the final section of the canal

The eastern end of Britain's longest canal tunnel, passing under Standedge Fell. Boats are now hauled through by electric tug.

The Lancaster Canal in Lancaster.

The Anatomy of Canals

Boaters emerge from the western end of Standedge tunnel to discover the splendid Pennine scenery around Diggle.

becomes increasingly urban as it approaches Ashton-under-Lyne. As a footnote to this brief look at the two Pennine canals: I first wrote about these two canals back in 1974 when I gave my opinion that both the Rochdale and the Huddersfield were full of interest, but were never likely to be restored. It just goes to show that canal historians should stick to looking back to the past and not try to forecast the future!

One other canal with links to Manchester was the Manchester, Bolton & Bury. This has remained virtually unused for half a century and more, but there are now plans for restoration being very actively discussed, so it seems sensible to leave discussion until the next volume, which will be bringing the canal story more or less up to date. Another new scheme will help to revive the fortunes of a canal which now exists in isolation from the rest of the system. The Lancaster originally ran from Kendal to Wigan, and part of it was used by the Leeds & Liverpool and then eventually bought out by the latter (Vol. 1). Over the years, there were further reductions: the link to the south was lost, and the northern half was severed when the M6 was built right across the line on an embankment. What remains is a twenty-four mile stretch of canal, between Ashton basin in Preston and Borwick. This part of the main line is lock-free, but a later arm down to the coast to Glasson has a modest six locks. Fortunately, the surviving section does contain one of Britain's most impressive aqueducts, crossing the Lune at Lancaster. The engineer was John Rennie, and as we saw on the Kennet & Avon, he was a man with a sense of style. At Dundas he designed a formal, classical structure in Bath stone. The architectural language used on the Lune aqueduct is the same, but the treatment creates a very different impression. This is an immense structure, 640ft long, crossing the river on five semi-circular arches, at a height of 60ft above the river bed. Lancaster is very much a northern city, built of local stone, not much given to fancy embellishments. So although Rennie incorporated classical columns and balustrades into his design, the stone surface is rusticated, squared off but not smoothed down. The result is a combination of Georgian elegance with northern roughness. In order to build the arches, coffer dams had to be made on the riverbed, and at the height of the work there were eighty men employed in double shifts, working day and night, just on pile driving alone. No expense was spared, whether the cost arose from importing special waterproof mortar from Italy for the foundations or supplying an extra beer ration to the men – 'to stimulate exertions' as the official reports recorded.

In order to achieve such a long pound, Rennie was forced to take a wide sweeping curve out of Preston, heading west for over three miles, then looping back again, before settling down for a much more regular line to the north. There are a number of small aqueducts along the way and one very notable deep cutting, just south of Lancaster. Burrow Heights is a hill rising just over 100ft above canal level, but it was enough to force Rennie to opt for a deep cutting, stretching for almost two miles. He certainly had no intention of climbing to a higher level, as he knew he would soon be faced with the Lune crossing: the aqueduct was quite high enough already. Beyond Lancaster, the going is comparatively easy at first, skirting the shoreline of Morecambe Bay, but up ahead there are sterner challenges as the hills build up towards Kendal and the Lakeland Fells. And it is at this point that everything is brought to a halt by the motorway. It is still possible to follow the line, and admire the excellent quality of the stonework in the locks, and peer into the 377 yard long Hincastle Tunnel. There are real hopes that the canal will not remain derelict for ever. In the meantime, work is well advanced on the construction of a brand new connection to the south, the Ribble Link, joining the canal to the Ribble and from there to the Douglas and the Leeds & Liverpool, ending years of isolation.

The Lancaster Canal at Preston is lined by cotton mills, many of which are immense structures built of local brick

The Anatomy of Canals

No. 1 Lock on the Glasson Arm of the Lancaster Canal, crossed by a bridge which demonstrates the elegance of John Rennie's work.

A major aqueduct carries that now disused section of the Lancaster canal over a minor road at Stainton.

The disused Lancaster Canal winding round the foot of Farleton Fell.

One of John Rennie's masterpieces, combining engineering daring with architectural finesse: the aqueduct over the Lune at Lancaster.

The Glasson Arm is a latecomer, only opened in 1825. It runs through some very open country and is as well constructed as the main line. The stone bridges are neat and tidy and include a particularly good stone skew bridge, number 8. Glasson itself has a large harbour, connected to the sea by a tidal lock. This was used until quite recently by quite sizeable coasters, but like many small British harbours is now generally given over to yachts and dinghies.

John Rennie made one other contribution to the canals of the north west, when he planned the Ulverston Canal. This is one of Britain's shortest canals, a mere mile and a half long, though when work began in 1793 there was talk of making a connection with the Lancaster. But even without the connection it was immensely successful for a time. It allowed ships to run up to the centre of the town, through an entrance lock 95ft by 22ft, and half a century after it was completed over 500 ships a year were passing through. They took valuable cargoes of iron and copper ores and slate out, and brought coal and timber in. Then came the railways, and the canal withered and died, though the old basin survives as a feature overlooked by the Hoad Hill 'lighthouse'. It is perhaps fitting that a basin with no ships should be overseen by a lighthouse with no light, for this is a monument, a stone copy of the famous Eddystone light. The Ulverston Canal is significant in that it was one of the first of a new generation of ship canals.

8. The Ship Canals

It is a mark of the growing confidence of canal engineers at the end of the nineteenth century that they began to turn their attention to ever bolder schemes. The narrow canals still formed the core of the English canal system, but broad canals had also been built – and now it was time to think of canals that could accommodate ocean-going ships. The idea was not entirely new – the Exeter Ship Canal had been completed in the sixteenth century, but it was short and was really no more than a river navigation, built to bypass a troublesome weir. The new ship canals were to be on a much grander scale. As always, the engineers produced their estimates and swore to their accuracy, and as always, the investors believed them and dreamed of huge dividends. And as well as estimating the costs, the engineers gave equally confident forecasts of just how long the work would take to bring to a triumphant conclusion. The fact that costs always seemed to increase and openings were never on time seemed to worry no-one in the heady days of the canal mania. But seldom have bright prospects been more comprehensively dimmed than during the long saga of what began life as the Gloucester & Berkeley Canal.

Work got under way in 1793, and in its essentials this was a very straightforward canal, even if it was built to a very impressive scale. The original plan called for a route from Gloucester, heading south for just over eighteen miles to rejoin the Severn at Berkeley Pill. As it was to be built along the flat Severn plain, the only locks required were those joining the canal to the river at either end. It was, however, to be 70ft wide and 15ft deep. This was to allow the popular blunt-bowed sailing cargo ships of the day, known as Indiamen, to reach what was to be the new inland port of Gloucester. At this time, the Severn was a hugely important transport route, navigable over a much greater length than it is today. The typical trading vessels were the trows, sailing barges with characteristic transom sterns, slab sides and rounded bilges which, by the end of the eighteenth century, mostly carried a fore and aft rig. They traded up-river, even as far as Ironbridge. They were well adapted for dealing with the difficult waters of the Severn, but the greatest hazard for bigger ships lay with the shifting sands and shoals of the upper reaches of the tidal river. It was this problem that the new canal was intended to circumvent. The engineer in charge was Robert Mylne, a man who had made his reputation in water supply rather than navigable waterways. But even if he had been the ideal man for the job he had, like most engineers in those hectic years, limited time to spend on personally overseeing the project. So the work fell largely to the resident engineer who came with the unfortunate record of having already been dismissed from two schemes. He lasted six months and his replacement, James Dadford, fared little better. Attempts were made to speed up the work by introducing the first mechanical digger to be set to work on a canal, but it was not a success. Mylne was sacked, Jessop was called in to offer advice, but by 1799 the money had run out and a mere five miles of canal had been built. It only restarted in 1817, with the help of a government grant. A new engineer, John Upton, appeared but overall charge went to Thomas Telford and eventually, in 1827, the canal was opened.

The canal itself was built to rather more generous dimensions than originally intended, with an increase in the depth to 18ft. There were, however, to be further major changes at

Sharpness Dock has been extensively modernised over the years, and modern silos dwarf the older warehouses.

the two ends. The canal was, if anything, too successful. Quite early on in the construction it had been decided to make the southern junction with the Severn at Sharpness Point rather than Berkeley, but even then the locks were proving inadequate for the trade. A new dock and entrance locks were constructed in the 1870s, enabling larger vessels to use the facilities. To meet the increased traffic, Gloucester Docks was also transformed with a major warehouse building programme. The canal we see today shows a mixture of eighteenth and nineteenth century technology.

Sharpness is that canal rarity, an inland port still trading at the end of the twentieth century, even if in decline, and as a result modernisation over the years has seen the destruction of much of the old and replacement with more modern facilities. The most striking features are still the stone piers, stretching out into the river, opening like a funnel mouth to guide ships in towards the entrance lock with its immense gates. The lock, 320ft long and 60ft wide has a depth of 24ft over the sill, though there is also a smaller barge lock giving access to the river. Among the old buildings to survive are the stables, modest brick buildings beside the locks, which were home to the horses used for towing sailing vessels on the canal. The dock master's house still stands and so does one of the original nineteenth century warehouses. The latter is typical of its type, a multi-storey brick building with a regular façade of windows and loading bays creating a strong visual rhythm. What is not so obvious is that Sharpness is, like the older Shardlow and Stourport, a canal new town and the terraced houses built by the canal company have an oddly suburban look to them. The old is literally overshadowed by the new, with concrete silos built in the 1930s and as recently as 1970.

Moving up the canal brings the original entrance lock into view, now abandoned and lined with rusting bollards. A circular stone pier is all that remains of the railway bridge that crossed the canal at this point. This took the lines across to Wales, but in 1959 two vessels collided in fog, were swept against the piers and the bridge across the Severn was no more. It has not been rebuilt. This being a ship canal, intended for use by tall-masted vessels, the only possible design for carrying roads over the waterway was the swing bridge. These required – and still require – the attention of a bridge-keeper who had to be housed on the canalside. These houses are among the oddest of any canal. Some have been modified over the years, and one of the best preserved can be found at Splatt Bridge near Frampton-on-Severn. It is built on top of the canal bank, so that seen from the waterway it looks ridiculously small, but is actually two storeys when seen from the land side. What distinguishes these cottages is the adoption of the Greek Revival style by the architect, with Doric porticoes facing the canal. They are widely admired, but I find the basic cottages to be overwhelmed by the decorative additions, rather as if they had been given a television make-over by a trendy designer.

The canal served factories along the way – such as the former Cadbury Chocolate works, but pays no attention to villages and hamlets. There is a small flurry of activity at Saul, where there is a level junction with the Stroudwater Canal. The main purpose of the canal is not to allow vessels to stop along the way, but to have a steady voyage up to Gloucester. Here the approach to the city is lined by concerns such as timber yards which took advantage of water transport, but the real centre of interest lies with the docks themselves. Like Sharpness, these have grown with the years. The first basin was completed in 1810 and an inner basin added in 1848. The oldest warehouse, with its upper storeys carried over the towpath on pillars dates back to 1836, the rest have a real coherence, all dating from the late nineteenth century. The style is very much that of the older building at Sharpness, but on a very grand scale. Though it is not immediately obvious from the architecture, not all these buildings had the same

The lock in the foreground is on the Stroudwater Canal, which makes a junction with the Gloucester and Sharpness at Saul.

The Ship Canals

The construction of the Gloucester and Sharpness Canal created an inland port at Gloucester.

Sharpness entrance lock and the dockmaster's house.

function. There is a link here to the huge grain silos at Sharpness. Some of the warehouses were converted to flour mills, one of the key factors in keeping the canal itself busy with trade. Now the old buildings are finding new uses, as offices, housing and very appropriately as home to the National Waterways Museum. It is difficult to envisage now, but this was in its day an international port, and the little Mariners Chapel of 1849 met the spiritual needs of the visiting sailors. Other needs were supplied in the scatter of pubs round the area.

This was not the only ship canal to be started in 1793. One of the problems faced by shipping along the west coast of Scotland was the Kintyre peninsula, a long finger of land pointing down from Loch Fyne and passing outside the island of Arran. At the top end, only a narrow neck of land separates Loch Fyne in the east from the Sound of Jura in the west. By constructing a canal just nine miles long shipping could avoid the 130 mile voyage round the Mull of Kintyre. So the Crinan Canal was constructed between Crinan and Ardrishaig. But where the engineers of the Gloucester & Sharpness had a level plain on which to build their canal, the men of Crinan were faced with a stony ridge in the middle which had to be overcome by locks. So, in this one short canal, there are sea locks at either end and a further thirteen locks to get over the summit, 59ft above sea level. A familiar story now unfolds, of a scheme begun in high optimism, which as difficulties multiplied ran out of cash, at which point the original subscribers simply walked away and refused to pay any more, and government money had to be used to complete the work. It was eventually opened in 1809, but not before stringent economies had been introduced, most notably reducing the depth of the canal from 15ft to 10ft at the western end. Unlike the delays on the Gloucester & Sharpness those on the Crinan had their origins in real engineering difficulties in driving the canal through the rugged, rocky landscape. However sensible it might have seemed at the time, the decision to reduce the canal's depth had unfortunate results as far as the canal's usefulness was concerned. Even the go-anywhere Clyde Puffers that carried so much of the trade of the Highlands and Islands in the early years of the twentieth century sometimes struggled to get through. Having made the journey myself I can still remember the ominous noise as the Puffer *Vic 32* scraped its way down the Crinan. The locks along the way are a comparatively modest 88ft by 20ft, which may have qualified the Crinan as a ship canal in the 1790s but is hardly generous. Add to this the fact that the locks were operated manually as on any inland canal, and that progress is further hampered by swing bridges, and it is not difficult to see why the Crinan was never a great financial success.

The harbour at Crinan itself is a delightful spot tucked in under the hillside, though the entrance from the sea has never been easy. The twin sea locks open straight out onto deep water with only the minimum protection from a short jetty. The first part of the canal follows a comparatively simple route along the edge of the estuary, with a heather covered hill, dotted by rocky outcrops to one side, and the broad sweep of mud flats on the other. Swing bridges provide the occasional pause before the climb to the summit begins. Looking at the water lapping against the bare rock at the edge of the canal, it is easy to see why the builders found life so difficult. The construction of lock chambers in particular involved a heavy programme of blasting and clearing. At Dunardry, one of the locks is crossed by a most unusual bridge, with a wooden platform braced by rods attached to pyramidal, lattice-work towers. It is moved by rolling along rails. At the opposite end, Lochgilphead and Ardrishaig still have evidence of trade with a large timber wharf. Here protective piers make access to and from the sea lock comparatively simple. Scotland's first ship canal was soon to be followed by a far more ambitious project.

The swing bridges that allowed tall ships to pass up the Gloucester and Sharpness were all manned. This bridge keeper's cottage has been extended.

The Caledonian Canal was first suggested in the 1770s. The journey around the north of Scotland has always been hazardous, and there is a tempting fault line through the centre of the Highlands, known as the Great Glen, already home to a number of deep lochs. A canal along this line would offer an easy, short passage from coast to coast. The route was surveyed in 1793 by John Rennie, but nothing was begun until 1802 when the job of chief engineer fell to Thomas Telford, though William Jessop was also deeply involved in the planning. There was no question here of running out of funds, for the scheme was backed by the government who saw the canal as fulfilling three valuable functions: bringing much needed improved transport and trade to the Highlands, giving work to the people of the region and providing a safe route for naval vessels. The canal is sixty miles long, but not all of it is artificial. It begins in the west with the sea loch, Loch Linnhe, then incorporates Lochs Lochy, Oich, Ness and Dochfour before joining the Beauly Firth in the east. Even though much of the canal uses these natural waters, the engineering work needed was still on a massive scale, unmatched by any other British waterway of the time.

Everything on the canal was big, with locks designed to take what seemed at the time very grand ships indeed – 150ft long by 35ft beam. As a result, there was no question of adopting the Crinan notion of having manually operated locks, worked simply by pushing on a balance beam. Instead the heavy gates were opened and closed using capstans, each one of which could be worked by up to four men. Similarly the swing bridges divide in the middle to make for easier movement. The bridge keeper would open one half, then row across to open the other. Today, mechanisation has removed the hard labour. The start in the west is spectacular, with hunch-shouldered Ben Nevis looming over the waterway. The first lock pokes out into the deep water of the sea loch, and a small lighthouse stands at the end of the pier. After that

The old Clyde puffer VIC32 passing through one of the narrowest sections of the Crinan Canal.

there is the canal basin, then the next two locks that begin the climb towards the next loch. The first swing bridges appear, one of them a relative newcomer, carrying the West Highland Railway which opened for business in 1901. Then the climb begins in earnest, through the eight inter-connected locks, known as Neptune's Staircase. It was considered a marvel in its day, but oddly does not look as impressive as other lock staircases, such as Bingley on the Leeds & Liverpool or Foxton on the Grand Union. The latter climb steeply and you get a real sense of the canal making a giant leap, but the sheer size of the Caledonian locks means that this staircase covers about a quarter of a mile, reducing the visual impact. There is a typical Telford touch in a lock cottage half way up, with a bow front, reminiscent of his designs on the Ellesmere Canal.

Having climbed 60ft, the difficulties were far from over. The canal follows the contours high above the wavering line of the River Lochy, and tributaries are crossed on two aqueducts. The larger of the two crosses the Shengain Burn on three stone arches. Even culverts and overspill weirs are built to a gargantuan scale. Loch Lochy is approached by two more locks in what seems a simple exercise in engineering. Look to the south, however, and a different story is told. The River Lochy has actually been diverted, and now leaves the loch in an artificial cutting below a dam before rejoining its old course. The dam was needed in order to raise the level of water in the loch to make it navigable. Telford then used the original riverbed for the course of his artificial canal. In the graphic phrase of his friend, the poet Robert Southey: 'Here we see the powers of nature brought to act upon a great scale, in subservience to the purposes of man: one river created, another (and that a huge mountain stream) shouldered out of its place, and art and order assuming a character of sublimity'. Thanks to Southey we have an eye witness account of the canal when construction work was at its busiest. Here there was more than the familiar army of navvies. Steam power had been brought in to pump water from the site where the lock staircases were under construction, and steam dredgers puffed away clearing a channel through the loch shallows. The scale of works looked forward to the scenes of the railway age.

The arrangements at Gairlochy by the entrance to the loch have changed since Telford's day. In 1834, floods sent water pouring right over the tops of the locks, so in 1844 a new lock was added to help control the flow. Even then it proved insufficient, and the lock gates were raised again in 1875 and a flood barrier created behind the lock cottage.

Only a mile and a half separates Loch Oich from Loch Lochy, yet that short route proved one of the most difficult to construct. Two locks lead up from the loch, but after that Telford

The Caledonian Canal has a dramatic beginning at Corpach, overlooked by the towering bulk of Ben Nevis.

The bow fronted house is a typical Telford design, and looks out over Neptune's Staircase at Banavie.

opted for a deep cutting to complete the route to the next loch. This involved blasting a way through the solid rock, and just how much material had to be removed can be seen in the vast spoil banks lining the canal. Little Loch Oich required yet more extensive dredging, particularly as this marks the summit with water draining away at both ends. The navigational channel is still comparatively narrow, marked by buoys. No doubt to the relief of the engineers, the next section of artificial cutting proved comparatively uneventful. There is one interesting point at Kytra lock where there was a bed of stone ideally situated to form the base of the lock, which saved the trouble and expense of building the usual invert arch. The cottage here is particularly charming in a traditional Scottish style with neat dormer windows. The calm ends at Fort Augustus with the drop down to Loch Ness. A five-lock staircase plunges down through the town and has become something of a tourist attraction.

Loch Ness, long and deep, provides an ideal route, the only problems occurring in stormy weather when it can seem very much like being at sea. The final section of wholly artificial cutting follows a line to the north of the River Ness as far as the outskirts of Inverness, where it swings north to make its own route to the sea. Now there was one final obstacle to overcome. The waters of the Firth are very shallow close to shore, so if the sea could not reach the canal, the canal would have to go out to reach the sea. Clay pits were opened up in a nearby hill, and the material used to build an embankment out from the shore. When it was high enough, stones were laid on top and given six months to settle. Only then could the canal be cut into the top of the bank and the sea lock built. The canal opened in 1822, well behind schedule and costing £900,000 instead of the estimated £350,000. In fairness, it has to be said that most of the costs were for labour, incurred because the scheme used untrained local Highlanders instead of the professional navvies employed elsewhere. This was deliberate policy, for providing work for the locals had been an important factor in the

The artificial nature of the Caledonian Canal is easily seen here, with the River Oich in the foreground and the canal at a higher level above it.

The lock cottage at Kytra on the Caledonian has clearly been derived from Scottish vernacular achitecture.

The very imposing lock staircase at Fort Augustus.

Laggan cutting represented a major achievement in terms of civil engineering, but is dwarfed by the magnitude of the Highland scenery.

The swing bridge at Moy had to be divided because of the width of the Caledonian Canal and the bridge keeper has to boat across to work it.

decision of the government to finance the scheme in the first place. It is still a magnificent waterway to travel. But although it achieved one of the aims in providing employment, it never met its other stated objective of providing a route for the Royal Navy in times of war. By the time it was finished, Britain was at peace and when naval battles again became a reality the ships had far outgrown Scotland's principal ship canal. The last words on the canal went to Robert Southey who wrote a poem in praise of Telford and his works, which was engraved on a wall at Clachnaharry. It is groaningly stilted: the opening gives the flavour of the whole.

Where these capacious basins, by the laws
Of the subjacent element, receive
The Ship, descending or upraised, eight times,
From stage to stage with unfelt agency ...

Enough! Telford has no need for such words, the canal speaks far more eloquently by itself.

9. Lost & Found

Many canals were built to perform a specific function, and when that function was no longer valid they were allowed to fall into dereliction. The Nutbrook Canal, briefly mentioned in the previous volume, is a case in point. It was built to serve the collieries of Shipley, in the heart of D.H. Lawrence country, very near his home town of Eastwood. It was entirely paid for by the colliery owners, to take their coal to the Erewash and it succeeded. It was a busy waterway, serving other collieries in the area as well via a system of tramroads, often laid on a temporary basis. It was assured of a good trade as there was also a major customer at New Stanton, close by the junction with the Erewash. This was, and is, an important iron making region, and the 4½ mile long canal thrived, but a glance at the modern Ordnance Survey map of the area is sufficient to tell the story of what happened. The whole area is covered by a network of railways, sprouting industrial branch lines, with one of the densest networks centred on New Stanton. At the end of the line, it can be seen not only that the mines have gone, but have been so completely obliterated that the colliery site at Shipley itself is now a country park. There was nothing left to carry by water. If one were looking for a canal to restore in order to explore the attractive surroundings along the way, this would not be it – but if you wanted to show how canals served industry this would definitely be a candidate.

The one part of the canal system that has survived and provides a popular amenity is the water supply system. Mapperley Reservoir has become a fishing lake, and is known to anglers for a record-breaking carp once caught here. Of the canal itself, there is still one section to be seen at New Stanton. The area is more notable for the sidings that took away the canal trade, spread out over a dusty wasteland, now that the works they once served have mostly gone. The steel pipe works at Stanton survive, and it is here that the canal emerges from a culvert to provide a hunting ground for more anglers, the former locks reduced to concrete waterfalls. Among the memories of the past are bars of pig iron still to be seen embedded in the ground near the canal. There is a tendency to mourn the decay of the canal system, but no one should feel sorry at the passing of the Nutbrook Canal. It did precisely the job it was built to do and did it well for just as long as it was needed.

The story of the Derby Canal is essentially similar, but it was a far larger concern. It ran from Swarkestone to Little Eaton, north of Derby, with a branch to the Erewash at Sandiacre. It was a straightforward enterprise with nine locks on the main canal and branch, and a further short branch with one lock to join the canal to the Derwent at Derby. It was, in many ways, a revolutionary canal. The engineer Benjamin Outram had apparently originally thought of continuing the canal north from Little Eaton as far as Denby, but his partner in the Butterley Iron Works, William Jessop, recommended instead that the steep climb would be better managed by means of a tramway. He then went further, by proposing an early version of containerisation. He suggested that as the loaded waggons would only travel downhill, one horse could pull a two ton load, and 'those Waggons may be drawn on to Boats and conveyed to Derby, and may be so constructed as to be carried into the Town without unloading'. The advice was followed, though not quite in the way Jessop had suggested. The coal was actually loaded into boxes, which were then lifted from the waggons and loaded into the waiting boats. The system continued in use for a remarkably long time, long enough in fact to be photographed. The pictures show a complex of lines, consisting of the typical arrangement of short lengths of cast iron plateway carried on stone sleeper blocks. The waggons appear to have been drawn in trains by teams of horses, and the boxes were lifted into place by a hand crane. There are few traces of the actual canal at Little Eaton, apart from a warehouse with a clock tower, but traces of the old stone sleeper blocks can still be found embedded in the ground.

Outram's other innovation has, alas, gone for ever. He had to take his canal over the Derwent on an aqueduct, to be built at The Holmes in Derby. As an ironmaster himself, it is perhaps not surprising that he proposed building it as an iron trough, and here we reach interesting and controversial territory. Outram began work at the same time as Telford and Reynolds were constructing their aqueduct over the Tern. Did they know about each other's work? Almost certainly, as there was a link in Jessop, Outram's partner at Butterley and Telford's superior on the Ellesmere. But who had the idea first and can claim the credit for innovation? That question is now unanswerable. Sadly, we can no longer see Outram's work. There is a description and photograph in Frank Nixon's *The Industrial Archaeology of Derbyshire*, published in 1969. It was made out of cast iron sections bolted together, each 6ft deep, 8ft at one edge, 9ft at the other, so that they could be fitted together at angled joints. Some time after that date it was dismantled and the parts sent for storage with the idea that it would eventually be re-erected in a museum setting. The ironwork was stored by the Highways Department but someone, it seems, decided it was a load of old scrap metal and out it went, all but two sections. It is a sad story, for this really can claim to be the first, if only in the sense that it was open to traffic a few days before the larger aqueduct on the Shrewsbury Canal.

The locks of the disused Nutbrook Canal have been cascaded, giving it an ornamental appearance, but making it unusable.

Looking down the Swinton locks to Waddington's Yard.

The one clearly defined section of the Derby Canal can be found at Swarkestone Junction. Here is a wharf with a simple wharf cottage, and a high arched bridge over the opening of the canal. A short length remains in water and is used for moorings, but beyond that there is only the towpath which has survived as a right of way for walkers and cyclists. It continues for approximately a mile before turning east under a disused railway to reach what was once the first of the locks – and the name Shelton Lock still survives in the built-up area where the canal now vanishes from sight. There are, however, plans to restore the line of the canal in the east from Sandiacre to Spondon on the outskirts of Derby, restoring features such as locks, but not putting the canal back in water.

Travelling further north brings us to what was once another important coal field in South Yorkshire and two more canals that were equally important in their day. The first of these is the Dearne and Dove where there is still, happily, a great deal to be seen. The canal begins on the navigable River Don at Swinton. The Don itself has been modernised with a neat and tidy control cabin at the lockside, with traffic movement managed by lights. Turning off onto the canal is a rather different experience. Six locks climb straight up from the junction, broad locks, 15ft wide but only 58ft long. These are the first of what were nineteen locks in the 9½ mile journey to Barnsley junction. There are two branches along the way, one of a little over two miles to Elsecar, with a further six locks and another of the same length, lock-free to Worsbrough.

Swinton has been home to the boat builders and carriers Waddingtons for a couple of centuries, and at first glance it looks as if they are still trying to get themselves organised. Barges lie around the basin in what seems a wholly haphazard manner and the collection of sheds around the dry dock seems to have been assembled with no clearer pattern in mind.

The locks on the Dearne and Dove are broad but short, so that the canal was never available for narrow boat traffic.

An artist's impression of the working days of the Dearne & Dove on a pub sign.

But as the company has survived so long it is safe to assume that they at least know what they are doing, even if it is a mystery to everyone else. The flight itself is interesting. The pounds between the big locks are relatively short, so to ensure that they could contain enough water to fill the lower locks, they were built with curved sides so as to be almost circular. This busy spot warranted the opening of no less than three pubs: the Towpath, the Ship and what was once the Red House. The Towpath acknowledges the past with a sign showing a horse pulling a coal barge. After this promising start it is disappointing to cross the road and find the canal has disappeared altogether, but it does re-emerge briefly among the spoil heaps of Wombwell. It is ironic that coal mining that brought prosperity to the canal for a time also killed it off. Subsidence caused severe problems, and as traffic dwindled, the company simply gave up maintenance and allowed it to fall into disuse.

This is where the first of the branches leaves the main line, and one can at least see how well this canal was built, for the surviving bridges are solid works, constructed out of rough hewn stone blocks. The most interesting part of this Elsecar branch appears at the far end at what is now the Elsecar Heritage Centre. It is based on the colliery and ironworks owned by Earl Fitzwilliam. He was one of the proprietors of the canal and its construction encouraged him to invest more money in the mines. In 1795 he paid for a new steam engine to pump water from the pit, but instead of turning to the 'modern' beam engines of the type now being supplied by Boulton and Watt, he opted for the earlier type of atmospheric or Newcomen engine. This is the last of its kind still to be seen on its original site, a massive

The Derby Canal has now been reduced to little more than a parking space for boats.

machine in a solid stone engine house. The atmospheric engines were slow and ponderous and used far more fuel than the new reciprocating engines – but fuel costs were not a problem for an engine set in the middle of a colliery.

The second branch is still very easy to trace along the river valley to Worsbrough. The branch also acted as a feed, supplying water from Worsbrough reservoir. The canal company reached agreement with the local mill owner that the same water could be used for his grain mill. The increased population, caused in good part by the improved transport system helping the growth of local industry, made for ever increasing demands for flour. So in 1843, the mill was extended and a steam engine added to the water wheel. Times change, however: the industry has shrunk, the mill is now a popular attraction and the canal has long since been abandoned.

The northern end of the Dearne & Dove brings a connection with the Barnsley Canal. This was promoted by the Aire & Calder Navigation, a very successful and busy company and anxious to expand connecting links to the industrial areas of South Yorkshire. The engineer in charge was the very busy William Jessop, and although the canal is now largely forgotten, it was by no means the least significant of his works. As one would expect, this is once again a broad canal, but with longer locks at 78ft 6in. It required a good deal of engineering works, with fifteen locks in 11 ½ miles. It had to cross the Dearne valley, and Jessop's original idea was to build an embankment and to cross the river with a single arch. But he was always a thorough man, and after testing the ground found that there was a good solid foundation of rock to either side of the river and elected for a five-arch aqueduct instead. Sadly it has been demolished, but some of the other impressive works can still be seen.

The Rutland Arms on the Grantham Canal, known to generations of working boatmen as the Dirty Duck.

The Barnsley Canal may have served a purely industrial world, but long sections of it have a very rural environment.

A deep cutting on the Barnsley Canal, comparable in scale to Jessop's other great cutting at Tring.

The line of the canal can be picked up north of Barnsley on the approach to Royston: it is not a very encouraging scene, but one can at least see the community the canal was built to serve, for mining remains can be seen on every side, both deep mines and later drift mines. The significance of the canal seems to have been lost on the owners of the Ship which, unlike the canal pub at Swinton, has opted for a fully rigged ship for its sign: few of these were ever seen in Royston! Nearby Royston Bridge Wharf offers a somewhat surprisingly rural theme. There were coal staithes here, but the house built right up against the canal bank still carries the word 'Dairy' carved in the stone lintel of the basement door. Travelling north marks a real change, for the canal comes out into the country, to be confronted by high ground which is sliced open by a deep cutting. In among the trees that line the steep banks are tumbled sandstone blocks, blasted away during construction. Much of the spoil was taken away and used for the next embankment. This summit level was supplied with water from Wintersett Reservoir, and extra supplies were assured with the construction of the Cold Hendley Reservoir, lying right alongside the canal. Here the canal wriggles round and through a hummocky landscape as the descent to the Aire & Calder begins with fifteen locks in less than three miles. Like all Jessop's canals it is carefully planned to minimise heavy engineering works and conserve water resources. The locks are all grouped together, so that once the top lock has been reached, there is a clear run all the way to the Dearne & Dove. The more one sees of Jessop's canals, the more one's admiration of this modest man grows.

The canals we have looked at so far in this chapter always made sense, connecting busy waterways that had already proved their worth to rapidly developing industrial regions. The same can certainly not be said about the next canal. The story begins with the construction of the Melton Mowbray Navigation, which was in effect a branch of the Leicester Navigation, constructed by making the Rivers Wreak and Eye navigable. It was a modest affair, a mere eleven miles long and serving a largely rural area. As soon as work began in 1791, a movement was started for an extension to Oakham. A glance at the map shows the area through which the canal was to pass was almost entirely rural, but no one it seems bothered to ask the basic question: where was the cargo to come from? There was vague talk about 'the produce of the country', but this was not a commodity that ever sustained trade on any canal on its own. Nevertheless, the work went ahead and even when the money ran out with only ten miles completed, they applied for a second Act in 1800 to raise more cash. It was finished, but was to enjoy only the briefest of histories. With such a fragile financial base it was always likely to prove susceptible to railway competition, but the proprietors were not to know just how aggressive that competition was to prove. The Oakham lay right on the line that the Midland Railway was contemplating for their new route, so they simply bought up the canal and closed it down after less than half a century of use. There is an interesting sidelight to this story. One of the main proponents of the Melton Mowbray and Oakham projects was Lord Harborough of Stapleford Court. He was not at all pleased with the Midland, and when they attempted to take a line through his parkland, he opposed them in every way. It was to result in a famous battle, between the railway surveyors, backed by navvies on the one hand and the gamekeepers and estate workers on the other. It ended with six railway men sent to gaol and a victory for the landowner. A wide detour had to be made, which for years was known to railwaymen as Lord Harborough's Curve. The canal enthusiast had a revenge of sorts, but far too late for the Oakham. There was, however, one group who profited from the demise of the canal, the shareholders. Having seen no income from the waterway in its years of trading, they finally made some money from its sale.

Jessop was involved in the early planning stages, though he left the actual construction to others, but it was his proposed route that was broadly adopted. The canal was to travel due north on the level for nearly seven miles to Edmondthorpe. Then it turned west to join the Eye valley for the line to Melton Mowbray with a total of nineteen locks in the remaining eight miles. It was this last section that had caught the attention of the railway engineers. It might be thought that, given its history, very little would have survived, and certainly in the Eye valley it has been more or less totally obliterated. Elsewhere there are traces to be seen, though it has to be said that, with this section being lock-free, there is little to find other than reedy channels, turning up in isolated sections. It can be traced from the edge of Oakham, but features are few and far between, even the old canal bridges having been demolished and replaced by concrete slabs. There is, however, one very interesting group of buildings at Market Overton. Here there was a very extensive wharf, with cottages and warehouses. The company may have been short of cash, but they seemed to have sufficient to spare to indulge in a little mild architectural embellishment. The stone used was a good quality limestone, and the drip mouldings above the windows give them a vaguely Tudor air. Behind the cottages was the wharf itself, approached from the road past a toll house or wharf office. Further along the road are two ranges of warehouses, now converted into private housing, though the regular pattern of small windows tells the obvious story. Nearby is a mile post, informing boatmen that they were now seven and a half miles from Oakham and eight miles from Melton Mowbray. But as far as the canal is concerned, this is the end of even this faint watery line, as a railway embankment rises over the bed of the Oakham Canal.

Is there anything to be gained from hunting down these memories of what was after all a notoriously unsuccessful canal? I think so. It was instructive to see just how dazzled presumably thoughtful business men had become by the excitement of the mania years. It just seems obvious with hindsight that there never could have been sufficient trade to show a meaningful profit, if any at all. It is equally interesting to discover that no one tried to do things on the cheap: these few surviving buildings show a concern for conscious design that is not always found even on far more successful ventures.

The next canal did at least have a clear objective, to reduce the transport costs to the important town of Grantham and supply places along the way with coal from the Nottingham Canal. Almost inevitably, Jessop was called in and early attempts to promote the Bill were lost due to the opposition of local mill owners. They feared that water drawn off from the Rivers Devon and Witham would have a serious effect on their businesses. Jessop rethought his plans and set out what was then a brand new idea, which he explained in a treatise, *Observations on the Use of Reservoirs for Flood Waters*. The idea that flood water alone would be sufficient to meet the needs of the canal was novel and important and proved effective, and the two reservoirs at Knipton (60 acres) and Denton (20 acres) did the job. There was, indeed, quite a job to be done, for the canal was built with broad locks able to take barges or a pair of narrow boats, and there were eighteen of them in the thirty-three mile length of the canal. Jessop opted for the obvious route down the Vale of Belvoir with a junction on the Trent nearly opposite that of the Nottingham Canal. The route is notably wayward, with Jessop always opting for the easy line rather than shortening the line through expensive earthworks. One happy result was that the canal which was authorised in 1793 was opened in 1796, and for once there would have been no need to go back to Parliament to ask for extra funds if it had not been for troublesome shareholders. Many who invested money in the heat of enthusiasm dreamed of fat profits and never bothered to read the fine

A mile post on the Oakham Canal.

print. The Act allowed the proprietors to make a call on the shares for extra funds up to an agreed amount. Those who had been prepared to pay once were less enthusiastic about doing so twice. It required a second Act to enforce the clauses of the first by imposing penalties for non-payment. So the canal was completed and for a time it prospered. Inevitably, competition arrived in the form of a railway from Grantham to Nottingham, but at least the canal owners could feel happy that they had negotiated good terms to compensate them for loss of trade. Alas, optimism was short lived. Railway companies were no more immune from money problems than canal companies had been before them, and when the time came to pay over the compensation, the coffers were empty. The canal had a brief spurt of activity during the First World War, when it was used to move army supplies, but then slipped back into the old pattern of gentle decline. By 1924 traffic had come to a halt, and in 1936 the inevitable closure was officially proclaimed. But if the canal now no longer answered its original function, it had, thanks to Jessop's planning, still a valuable role to play in the land. It was still a means for collecting and delivering flood water. So the line remained more or less intact, though the locks, which were no longer needed, were cascaded. Bridges were dropped, cuts through embankments culverted. Nevertheless, the waterway itself remained, clear and deep – and walking along the former towpath today it often seems in better condition than several alleged navigable waterways one could name. It is not surprising that restoration work is already well under way.

One of the few difficult sections occurred right at the beginning of the route out of Grantham. The country is knobbly rather than actually hilly, but it still involved a good deal of twisting and turning to keep a level. The first part in the line has been lost under housing development, and now begins about a mile from the original terminus and is soon overwhelmed by the modern ring road. This was not here to bother Jessop and the engineers who were responsible for the actual construction, but they did have to confront Harlaxton hill. Here they opted for the straight line, going through in what is now an attractive, deep cutting. The first feature of note is the Denton reservoir, just to the south of the canal, beyond which the line becomes ever more wayward, laid out on the land as a giant letter M. The first of the flight of seven locks is reached at Woolsthorpe, a spot inevitably marked by a canalside pub – set alarmingly close to a spot called 'Brewer's Grave'. The dominant feature in the landscape is Belvoir Castle sat up on a wooded hillside and the owner, the Duke of Rutland, had a considerable say in the way the canal was built as it passed his lands. He was not averse to its presence, far from it, as it provided a convenient means of supplying the castle with necessities, especially coal. He was not, however, prepared to have anything that interfered with the appearance of his grounds. Jessop's second reservoir was to be built immediately to the south of the castle at Knipton, but the Duke insisted that the feeder had to be culverted at considerable expense, hidden from view until it reached the canal. The canal fraternity had their revenge in their own way: the Rutland Arms pub became for ever known as the Dirty Duck. One can have some sympathy with the Duke, for this is beautiful countryside, but the canal enhances rather than disturbs the scene. The next straight section seems to be blessed with an unusually wide towpath, but this in fact marks the line of an old ironstone railway which for a time kept the canal company. They diverge again near the very attractive brick bridge at Stenwith.

The canal appears briefly in water near Oakham, but is soon lost again in reedy shallows before disappearing altogether.

The Duke of Rutland's landscape concerns extended only as far as his own interests determined. It is difficult to think of a little feeder as an eyesore: certainly it would be more attractive than a railway. But the Duke's supplies needed to be carried up the hill to the castle. He had a private wharf built at Muston Gorse and from here a tramway swept up the hillside, the line still defined by a track. Railways again begin to dominate the canal, and it is easy to see just why they were a threat, as lines run alongside and over the waterway, promising a multitude of connections. Ironically, they have closed while the canal is still in water and sections are already back in use. One reason why the canal is in such good condition becomes clear at Harby. A small group of unadorned, very functional buildings represent the working maintenance yard, kept going to ensure the watercourse did remain clear. There are not many grand buildings along the way. This was always a rural canal, running through a thinly populated area. One of the first areas to be restored was the canal basin at Hickling. The basin itself is square with a wharf building hardly any grander than the average barn. In fact, it looks very like a barn with wide waggon doors at ground floor level with a loading bay above. It is an odd asymmetrical building of red brick with a pantile roof. The wharfinger's house is set further back and among the houses is the Plough Inn, suggesting that the locals still regarded farming as more important than the watery newcomer. The refurbishment of the area has included the instalment of a row of bollards, presumably in anticipation of boats again using the waterway. Unfortunately, they were put beside the road instead of by the canal. You could hitch up a bicycle, but if you tried to tie a boat, the rope would stretch straight across the footpath. Very odd.

The Nottingham end of the canal brings a change in character, with the first real indications of industry appearing. There were gypsum mines here in the nineteenth century, a

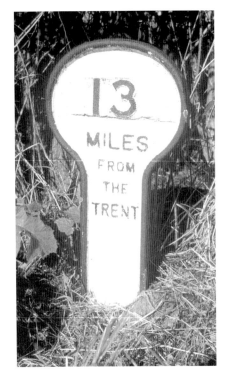

A cast iron milepost on the Grantham Canal.

material used for many things from making plaster of Paris to blackboard chalk. Unfortunately the strata under the canal made it susceptible to leakage, a problem which constantly tested the engineers' patience. Rather more familiar mining is also in evidence. On its way to join the Trent, the canal passes close to the big Capstone colliery. Lock restoration is already well under way in this section. As at the Grantham end, however, the Nottingham end has been lost under new buildings and a new junction must now be made. Although full restoration to navigation is a long-term project, an excellent start has been made on repairing the towpath so that it will be possible to walk the full length of what is a grand canal, passing through very attractive countryside.

Sometimes, it seems, a canal was hardly worth the work that was put into its construction. But in 1792 two gentlemen, Bruges and Miller, together with a Mrs Sarah Mumford put their resources together to build a canal. It was an extremely modest affair, built with the sole purpose of getting coal from the Severn to what was the growing and increasingly fashionable town of Cheltenham. It joined the river at Wainlode, where there was a double entrance lock, after which it headed off towards Cheltenham. Then, after less than three miles, the rising ground of Coombe Hill was reached and the canal came to an end, still five miles short of the town. Why they did not decide simply to build a road for the whole journey will be forever a mystery. It certainly puzzled the canal's various owners, several of whom went bankrupt in the attempt to run it at a profit. They were still pondering over closing it when the decision was taken out of their hands. In 1876 a major flood swept away the lock gates, damaged the locks and no one ever bothered to repair them. Surprisingly, however, the canal remains in water and a cluster of cottages and a wharf house still stand beside the reedy basin. The Coombe Canal, which never found much use as a commercial waterway, is enjoying a modest old age as a nature reserve.

The decade of the 1790s was above all a time of contrasts. In the furious spate of canal building that marked the period, there were canals which were to provide both vital links to the network as a whole and to prove profitable to their owners: and there were canals doomed to disaster from the first. The ingredients for success had never really changed from the days of the first canal: the route needed a good, bulk cargo that could provide a regular trade. Even with these elements in place, there was still the risk of failure if the necessary engineering works were too heavy, delaying opening and piling on the costs.

Of all the canals that got things wrong, none did so quite so spectacularly as the Hereford & Gloucester. It began, as ever, with high optimism in 1791 under the direction of Josiah Clowes, who had begun his career as assistant to Whitworth on the Thames & Severn. Things had scarcely got underway before everyone changed their minds about the direction the canal should take and had to get approval through a second Act in 1793. This is not as surprising as it seems, for this is very difficult country for canal builders, full of little hills and laced with rivers and streams. There is no obvious way through – and the route that was eventually decided upon is certainly not obvious either. Gloucester lies roughly twenty-four miles to the south-east of Hereford, but the canal was over thirty-five miles long and began by heading almost due north. The idea was to find a suitable spot to cross the River Lugg and to avoid the steep hills further south. So it was only after some three miles that a start was made in approximately the right direction for Gloucester. With a great deal of twists and turns, the first objective in the east was reached, the River Leadon, which was crossed. Now it was possible to head down the river valley to the south, but even then there were immense difficulties to be overcome. Normally one would think that by reaching the valley things would

Restoration is well under way on the Grantham Canal. This is Hickling basin with its unusual asymmetrical warehouse.

get simpler, but the canal is constantly hopping from one side to the other. Below Ledbury the line turned to the east for another crossing of the Leadon, and there were still another four aqueducts to come before reaching the Severn.

The wayward route might have been acceptable, but there were locks to be built as well, for the summit level was almost 200ft above the Severn. Worst of all were the tunnels, three of them in all, and not minor affairs either: one near Hereford a quarter of a mile long, a slightly shorter one at the summit and then the tunnel that swallowed up all the funds at Oxenhall, a daunting 2,192 yards. As the works inched forward there were many times when the proprietors contemplated abandoning the project and cutting their losses: they would have been well advised to do so. The canal only reached completion, after an amazing half a century of effort, in 1845. By then canals were already considered old-fashioned anachronisms. Railways were all the thing. By 1881 it was closed again, bought up by the railway interest, who used the bed for a line from Newent to Gloucester. The canal was never likely to be commercially viable, and certainly was unlikely to repay the expense of construction through such difficult countryside. Sections do remain in water between Hereford and Newent: at the latter place, by a neat touch of irony, the railway has in its turn been swallowed up by the Newent bypass. Serious work is already under way at the Gloucester end, where a canal cenntre has been established by the junction with the Severn.

The Ashby Canal, unlike the Hereford & Gloucester, got things right, though only because they changed their minds after the Act was obtained in 1794. The original plan was, as the name suggests, to make a canal through Ashby-de-la-Zouch. It was to start at the Coventry Canal and end at the limestone workings near Ticknall. One of the first things I do when looking at the line of a canal is go across to my map shelf and take down the appropriate

1:50,000 Ordnance Survey maps. Here you can read the nature of the countryside, see the dips and hollows, the hills and ridges. The canal engineer enjoyed no such luxury. He had to rely on his own surveys and often these were no more than reconnaissances in the limited time available between the birth of the idea and the application to Parliament. My map shows a mass of contour lines on the approach to Ashby, and no respite beyond the town. A thorough study of the ground by the surveyors rapidly led to the conclusion that canal building on this terrain would be difficult and expensive. It was estimated that the final stretch would involve lockage to overcome a change of level of 252ft. The alternative was wonderfully attractive. End the canal at Moira, and the only lock required throughout the whole length would be the stop lock to control movement between the Ashby and Coventry Canals. There was still a need to feed in the valuable cargoes coal and limestone at the northern end, but that could be done much more efficiently by tramway than by water. The work went to Robert Whitworth and Jessop and together they devised a route that followed the 300ft contour – or as the modern map shows it the 100m line. It is worth taking time to look at the modern OS map to see just what the engineers did. Mostly where the contour turns, the canal turns, but occasionally the line is crossed and there you will find signs of earthworks, and at Snarestone the one tunnel on the line was constructed. The result of all these changes was that the canal that was once to be fifty miles long was reduced to thirty and was opened in 1804.

A lock-free canal is not necessarily an interest-free canal, as the Ashby soon demonstrates. Once past the stop lock at Marston junction, the colliery spoil heaps that had been such a marked feature on the Coventry Canal, give way to agricultural land and the canal soon dives down into a deep cutting. As if to point up the change of character between the immediate surroundings of the Coventry and the Ashby, on my last visit by boat, a kingfisher flashed to and fro across the water in a colourful flying display. It is very much the case that this end of the canal gives little hint of its former, busy industrial trade. Inevitably a canal that travels so far without locks has to take evasive action somewhere along the route and the Ashby includes a number of sharp curves. It also never quite reaches any of the towns and villages along the way, so that the engineers had to provide wharves at the nearest access points. The first town to be met is Hinckley which was one of the early centres for framework knitting. The frame was one of the first devices to be introduced into the textile world, to replace an old handcraft by a more efficient machine. It was designed as early as 1589 by a Nottinghamshire clergyman, William Lee. There is a charming, if unlikely, story that the young reverend was passionately in love but the lady was too engrossed in her knitting to pay attention to his wooing. So he determined that he would make a machine to leave future young women free to turn their attention to more important matters than clicking needles. The frame was used for making stockings, and though not immediately successful, by the time the canal came along there were thousands of frames at work in houses and workshops in Leicestershire and neighbouring Nottinghamshire. So, although the canal appears to go through a very rural part of the world, there were villages and towns all along the way where machines were busily at work. Today, the canal at Hinckley is mainly notable for its big, new marina, which is set to undergo more development.

The canal continues to go its own lonely route, with just the occasional road or accommodation bridge along the way. These are mainly typical canal bridges, but are notable for the very high quality of their stone work. At Stoke Golding, the canal again only reaches to within half a mile of the village centre, so the wharf provided a new nucleus for development. It has

The Ashby Canal is deceptively rural for most of its length, and was built without locks but with some extravagant bends.

now found a new use as home to a hire boat base. You can still see how the village expanded in a linear fashion from the old tight-knit centre down as far as the road bridge over the canal and then halted. After that the canal goes into a U-bend round a hill, and the next village is missed, literally, by a mile, so another lonely wharf had to be constructed to serve Sutton Cheney. Now the rural peace is occasionally disturbed by what is for some of us an enticing sound, the blast of a steam whistle. In 1873, the canal trade was threatened by competition from a line, also designed mainly to serve the area's collieries, running down from Ashby to a junction with the main line near Nuneaton. In fact, the canal has outlasted it quite comprehensively, for the route closed to passengers as early as 1931. Now a section has been restored for a steam passenger line, and after several name changes has settled down as The Battlefield Line. The name is appropriate, for its southern end is at the terminus by the Battle of Bosworth centre. This was the battle which changed English history, as Richard III famously lost his life offering his kingdom for a horse, and the first of the Tudors took the throne. The canal also skirts this most famous battlefield and provides a good view as it runs along the top of an embankment before crossing an aqueduct on the road to Market Bosworth.

Canal and railway remain close companions past Market Bosworth wharf and station to Shackerstone. This stretch gives one a chance to compare the different lines taken by engineers, working almost a century apart. The canal takes a very wriggly line, whereas the railway is altogether more direct and by the time Shackerstone station is reached, the railway is high above the older waterway, carried on a long embankment. Shackerstone station is very close to the canal, and it is worth calling in if only to see the attractive little museum set up by a former signalman John Jacques. It was a labour of love, quirky but delightful with every inch of space filled. Over to the east a long avenue of trees once led to Gopsall Park, the house where Handel stayed while writing the *Messiah*. The house has been demolished and the parkland has lost some of its well-groomed elegance. A measure of its former importance is a clause in the original Act that allowed for a fine of £50,000 if any water was abstracted from the estate or if anything was done to harm the grounds. Considering that the sum authorised for construction of the entire canal, which at that time was still expected to go all the way to Ashby, was only £150,000, this is a huge amount of money. The most notable feature in the landscape is now the low swell of hills across the route. These required the one major piece of engineering along the way, Snarestone Tunnel, 250 yards long, but with no towpath.

If the canal owed its prosperity to coal mining, there was also a price to be paid. Subsidence became an ever increasing problem. Gradually the northern end of the canal was abandoned, until eventually everything came to a halt a little way beyond the northern end of the tunnel. The final nine miles of canal were officially abandoned. Much of the towpath remained open so that it was possible to see how the land had dropped away, in effect creating embankments where none had ever been envisaged when the canal was new. Now the restorers have set to work, and the lock-less canal is lock-less no more – land level and canal level are again

The Ashby Canal has recently been reunited with its industrial heritage. This is the immense blast furnace built at Moira c.1800.

reunited. It is on this last section that the true, industrial character of this canal becomes evident for the first time. Measham has a large brickworks on the edge of the town, and local clay was also used in the famous pottery. In particular, this was the town that made so-called bargeware. They produced large teapots, which were notable for a second small pot as decoration on the lid. The name was acquired because boating families bought them on their way through and their popularity spread throughout the connected Midland system. Donisthorpe brings collieries, but by far the most impressive site is to be found at Moira. Here an immense blast furnace was built beside the canal. Iron ore is found around Ashby, coal mines are close at hand, and the canal could take away the finished product, so this was the logical place to site works in about 1800. It is one of the biggest blast furnaces to survive from the period, and one can still see the bridge along which the raw materials were brought to the top of the furnace, and the arch through which molten metal ran into the moulds on the casting house floor. The engine, which once blew air into the furnace, has long gone. The furnace has its own museum, and craft workshops have now been added.

Moira is a good place to sum up much that happened in the mania years – and a good deal of what has gone on since then. Here was a canal built to serve the great industrial expansion of the late eighteenth century. Once the decision to build had been taken, the proprietors turned for advice to the man who was to see so many of the canals of the age through to a successful conclusion – William Jessop. The canal prospered in its early years, but suffered competition in the railway age – and was eventually taken over by the railways, in this case the LMS. When troubles came in the form of subsidence, there was no great enthusiasm for remedial work. Then, after the Second World War when the commercial future of canals everywhere was being questioned, the decision was taken. The difficult part would simply be abandoned. Recent years have seen the new growth industry of the canals take over – pleasure boating, and now the abandoned section is again being brought slowly back to life. When I first travelled on the Ashby many years ago, I never expected to see restoration of the northern end – probably never even considered it. But then, other canals which seemed even less likely candidates for restoration have been fully restored in recent years. And now, an even less likely scenario has appeared – new canals under construction in the twenty-first century. The end of construction in the nineteenth century, and the great changes brought about in recent times will form the subject of the last volume.

Gazetteer

The following list gives the more interesting sites to be found on the canals covered in this book. The standard Ordnance Survey system of grid references is used with one small variation. The first number quoted gives the number of the Landranger (1:50,000) map on which the feature appears. This is followed by either a 4 or 6 figure number. The former gives a reference to features such as flights of locks and basins that are spread over a large area; the latter gives a more exact location to special features, such as bridges and warehouses.

Chapter 1
    Shropshire Canal   Hay incline, Blists Hill   127/695028
    Shrewsbury Canal   Longdon-on-Tern aqueduct   127/617157
    Worcester & Birmingham   Diglis Basin   150/850549
                                        Stoke Wharf   150/943663
                                        Tardebigge top lock   139/988689
                                        King's Norton junction   139/053795
                                        Bournville wharf   139/045827
                                        Gas Street Basin   139/0686
    Stratford Canal   Guillotine lock   139/054794
                                        Lapworth locks   139/1870
                                        Wootton Wawen aqueduct   151/159630
                                        Edstone aqueduct   151/162609
    Dudley No.2 Canal   Windmill End   139/953882

Chapter 2
    Grand Union Canal   Little Venice   176/2482
                                        Bull's Bridge   176/108791
                                        Three Bridges   176/143798
                                        Grove Park Bridge   176/087988
                                        Tring cutting   165/9313
                                        Wolverton aqueduct   152/801418
                                        Stoke Bruerne   152/7449
                                        Weedon barracks   152/6259
                                        Braunston junction   152/533660

Chapter 3

Grand Union Canal	Hatton locks	151/2466
	Camp Hill locks	139/0482
Leicester Arm	Watford staircase	152/5968
	Foxton staircase	141/6989
	Junction with Soar	140/567009

Chapter 4

Ellesmere Canal	Hurleston locks	118/6255
	Grindley Brook	117/5243
	Ellesmere Basin	126/3934
	Frankton junction	126/370319
	Chirk aqueduct	126/287371
	Pontcysyllte	117/271420
	Horseshoe Falls	125/196433
	Ellesmere Port	117/407770
Montgomery Canal	Carreghofa locks	126/254202
	Vyrnwy aqueduct	126/254195
	Berriew aqueduct	136/188006

Chapter 5

Monmouth Canal	Fourteen locks	171/269892
Brecon & Abergavenny Canal	Pontymoile junction	171/294003
	Ochram turn	161/2909
	Llanfoist wharf	161/285130
	Brynich aqueduct	160/079273
Glamorgan Canal	Merthyr Tydfil	160/044055
Neath & Tennant Canals	Aberdulais aqueduct	170/772993

Chapter 6

Kennet & Avon Canal	Sydney Gardens bridges	172/7565
	Claverton pumping station	172/791643
	Dundas aqueduct	172/785625
	Avoncliff aqueduct	173/803600
	Caen Hill locks	173/9861
	Crofton pumping station	174/262623
Somerset Coal Canal	Midford aqueduct	172/757605
	Combe Hay incline	172/742606
Basingstoke Canal	Deep Cut locks	186/9256
	Aqueduct across the railway	186/893565
	Greywell Tunnel	186/718514

Chapter 7

Ashton Canal	Ducie Street junction	109/849982
	Dukinfield junction	109/934982
Peak Forest Canal	Marple aqueduct and locks	109/955901
	Buxworth basin	110/025822
	Whaley Bridge interchange	110/013817
Rochdale Canal	Summit	109/9418
	Hebden Bridge basin	103/993271
	Sowerby Bridge	104/055239
Huddersfield Canal	Mills and reservoir	110/055125
	Standedge Tunnel end	110/040120
	Shaft and engine house	110/027105
	Tame aqueduct	109/9906
Lancaster Canal	Lune aqueduct	97/484639
	Glasson dock	102/4455
Ulverston Canal	Entrance lock	96/313777

Chapter 8

Gloucester & Sharpness Canal	Gloucester Docks	162/8218
	Splatt Bridge	162/742068
	Sharpness docks	162/6602
Crinan Canal	Crinan harbour	55/7894
	Dunardry bridge	55/820912
Caledonian Canal	Entrance lock, Corpach	41/098767
	Neptune's Staircase	41/1177
	Laggan cutting	34/2997
	Fort Augustus locks	34/3709
	Sea lock Clachnaharry	26/643468

Chapter 9

Derby Canal	Swarkestone junction	128/372292
Dearne & Dove Canal	Swinton locks	111/463990
	Worsbrough mill	111/350033
Barnsley Canal	Deep cutting, Royston	111/3613
Oakham Canal	The wharf, Market Overton	130/881161
Grantham Canal	Woolsthorpe locks	130/8435
	Hickling basin	129/691293
Coombe Canal	Terminus	162/887273
Ashby Canal	Shackerstone	140/3706
	Moira furnace	128/314152

Index

Nos in italics refer to black and white photographs

Abercraf,	81
Abercynon,	80
Aberdare,	80
Abergavenny,	76-7
Abingdon,	33, 84
Aldershot,	91
Alvechurch,	11, 15
Ancoats,	*96*
Apsley,	32
Aqueducts	
Aberbechan,	74
Aberdulais,	81
Avon, Warwicks,	49
Avoncliff,	83
Brynich,	79
Caerfanell,	79
Cam Brook,	88
Carreghofa,	71
Derby,	65, 131
Dundas,	83, 86, 110
Edstone,	20-1
Fray's River,	30
Great Ouse,	37
Hebden,	104, *105*
Lledan Brook,	73
Longdon on Tern,	9, 21, 65, 131, *8*
Luggy Brook,	74
Lune,	110-1, *114*
Marple,	98, *100*
North Circular Road,	29
Oxford & Birmingham Railway,	49
Pontcysyllte,	62, 65-6, 68, 67
Rea,	52
Rhiw,	74
Shengain Burn,	122
Stainton,	*112*
Tame,	96, 108
Vyrnwy,	72, *72*
Whitewater,	94
Wootton Wawen,	20
Yarningale,	20
Ardrishaig,	120
Arran,	120
Ashby-de-la-Zouch,	144, 145, 147, 148
Ashton-under-Lyne,	95, 99, 110
Atkins, Charlie,	17
Aylesbury,	33
Aylestone,	56
Barnsley,	138
Basins and wharves	
Ashton,	110
Buxworth,	99, *101*
Camp Hill,	52
Castle Foregate,	9
Coombeswood,	23
Devizes,	84
Diglis,	14
Droitwich,	15
Eagle Foundry, Leamington,	48
Frogmore,	30
Gas Street,	18
Goytre,	78
Hanbury,	15
Hickling,	142, *144*
Linford,	37
Llanfoist,	78-9, *79*
Lowesmoor,	14
Market Harborough,	56
Muston Gorse,	142
Pewsey,	84
Paddington,	26-7
Queen's Head,	71
Royston Bridge,	138
Stoke Works,	15
Stratford-upon-Avon,	21
Uxbridge,	30
Weston,	70
Basingstoke,	90, 94
Bath,	82
Bathampton,	83
Beaufort, Duke of,	81

Index

Beauly Firth,	121	Moy,	*129*
Beech House, Ellesmere,	65	North Warnborough,	94
Bele Brook,	72	Oich,	*125*
Belvoir Castle,	141	Pontypridd,	98
Belvoir, Vale,	139	Prees Arm,	*64*
Ben Nevis,	121	Sounding arch,	22
Berkeley,	116	Splatt,	118
Berkhamsted,	*27*	split, Stratford Canal,	20
Big Pit Mine,	79	Sun Trevor,	70
Birmingham,	10, 16, 17, 23, 44, 49, 52, *11*	West Drayton,	30
		Worcester railway,	14
Black Jack's Mill,	30, *31*	Bridgewater, Duke of, 81,	95
Blaenavon,	78-9	Bridgnorth,	61
Blists Hill,	8-9	Bristol,	82
Blorenge,	77, 78	British Waterways,	17
Boats		Briton Ferry,	80
Cressy,	69-70	Bruce, Thomas, Earl of Ailesbury,	85
Kildare,	39	Bruges, Mr.	143
Paddington packet,	30	Bull's Bridge,	29, 35, *28*
President,	39	Bull's Nose,	88
trows,	116	Burrow Heights,	111
tub boats,	7, 23	Butterley ironworks,	65, 98, 130, 131
VIC 32,	120, *122*	Buttington,	73
Borwick,	110		
Bosworth, Battle of,	146	Cadbury,	17, 118
Boulton & Watt,	85, 134	Canals	
Bourne, J.C.,	33	Aberdare,	80
Bournville,	17, 20	Aire & Calder,	136, 138
Bradford-on-Avon,	83, 86	Ashby,	144-8, *146, 147*
Braunston,	24, 30, 44, 45, 46-7, 52, *40, 43*	Ashton,	95-6, 99
		Barnsley,	136-8, *137*
Brecon,	76, 79, 80	Basingstoke,	89-94
Brecon Beacons National Park,	75	B.C.N.,	10, 17, 22
Brentford,	24, 29	Birmingham & Fazeley,	24, 45, 46, 52
Bridges,		Birmingham & Liverpool Junction,	69
Barley Mow, Winchfield,	91	Brecon & Abergavenny,	75, 76-80, 98, 78
Bascote,	47		
Braunston,	37, *38*	Bridgewater,	25, 99
Brick Kiln,	93	Calder & Hebble,	99, 104
Budbrooke,	48	Caledonian,	120-9,
Candle,	41-2	Cheshire Ring,	96, 101
Galgate,	*112*	Chester,	60, 62
Gibraltar,	47	Coombe,	143
Gorrell's,	103	Coventry,	24, 144, 145
Grove Park,	31, *32*	Crinan,	120, 121, *122*
Ironbridge,	9	Cromford,	24
March Barn,	103		

Dearne & Dove,	`133-6, *131, 133*	Regent's,	27
Derby,	130-3, *135, 136*	Rochdale,	95, 99-104, *105*, 110
Droitwich Junction,	15	Shrewsbury,	9, 65, 69
Dudley No.2,	17, 21-3	Shropshire,	7-9
		Shropshire Union,	7, 17, 60, 74
Ellesmere,	9, 60-62, 65, 71, 122, 131, *64, 70*	Soar,	53, 56
		Somerset Coal,	83, 86-9
Exeter,	116	Staffs & Worcester,	10, 76
Glamorgan,	40, 80	Stratford,	16, 18, 21, 52
Gloucester & Sharpness,	17, 116-120, *117, 121*	Stroudwater,	118
		Swansea,	80-1
Grand Junction,	24, 25-44, 45, 46, 52-3, *25, 26, 27, 28, 31, 32, 34, 35, 36, 38, 39, 40, 41, 42, 43, 46*	Tennant,	80-1
		Thames & Severn,	143
		Trent & Mersey,	9, 25, 106
Grand Union, see also Grand Junction, Warwick & Birmingham, Warwick & Napton;		Ulverston,	115
		Warwick & Birmingham,	20, 45-6, 49-52, *48, 49, 50, 51, 53*
Northampton Arm,	42, 52, *54*;	Warwick & Napton,	46-9, *47*
Leicester Arm,	44, 51-6, *55,56, 57, 58, 59*	Wilts & Berks,	84
		Worcester & Birmingham,	10-18, 21, 22
Grantham,	139-43, *136, 142*	Capstone Colliery,	143
Hereford & Gloucester,	143-4	Cardiff,	60, 80
Huddersfield,	96, 99, 101, 105-110, *106, 107, 108*	Carreghofa,	65
		Cassiobury Park,	31
Huddersfield Broad,	99, 105	Chapel-en-le-Frith,	98
Kennet Navigation,	86	Cheltenham,	143
Kennet & Avon,	82-6, 110, *85*	Chirk,	62, 66, 68
Ketley,	8	Clarendon, Earl of,	31
Lancaster,	110-5,	Claverton,	82
Leeds & Liverpool,	26, 99, 110, 115, 122	Coalbrookdale,	8, 71
		Coalport,	8
Leicester Navigation,	138	Combe Hay,	88
Leicester & Northampton Union,	52	Coombe Hill,	143
		Crawshay, Richard,	80
Macclesfield,	98	Crick,	54, *58*
Manchester, Bolton & Bury,	110	Crinan,	120
Melton Mowbray,	138	Crofton,	85, *88*
Monmouthshire,	62, 75-6	Cromford,	99
Montgomery,	69-74	Crookham,	91
Neath,	80	Crumlin,	75, 76
Nottingham,	139		
Nutbrook,	130, *131*	Darby, Abraham,	9
Oakham,	138-9, *140, 141*	Debdale,	52
Oxford,	24, 44, 45, 46	Deepcutting,	73
Peak Forest,	96-9	Denby,	130
Ribble Link,	111, 115	Derby,	130, 131

154

Index

De Salis, Henry, 89
Diggle, 108, *110*
Dobson's Bridge marina, 65
Docks
   Adelaide, 29
   Glasson, 110
   Gloucester, 118-21, *119*
   Sharpness, 117, 118, *117,119*
Dogmersfield, 91
Donisthorpe, 148

Ealing, 29
Eastwood, 130
Ebbw Vale, 75
Eddystone lighthouse, 115
Edgbaston, 14
Edmondthorpe, 139, *140*
Edwards, William, 98
Ellesmere, 60, 65
Ellesmere Port, 60, 69
Elsecar, 133, 134-6
Emscote, 49
Engine Wood, 88
Engineers
   Barnes, James, 24, 37, 52
   Brindley, James, 6, 9, 19, 25, 53, 76, 90, 106
   Brown, Thomas, 98
   Brunel, Isambard Kingdom, 27
   Buck, George S., 69, 71, 73, 74
   Clowes, Josiah, 9, 10, 13, 18-19, 143
   Dadford, James, 116
   Dadford, John, 69
   Dadford, Thomas Jnr., 75, 78
   Felkin, William, 50-1
   James, William, 18-19, 20-1
   Jessop, Josias, 69, 74
   Jessop, Wiliam, 6, 9, 24, 32-3, 37, 52-3, 61-2, 63, 65, 66, 82, 101-3, 116, 121, 130, 131, 136, 138, 139-41, 145, 148
   Mylne, Robert, 116
   Outram, Benjamin, 65, 98, 130
   Porter, Samuel, 18
   Rennie, John, 13, 15, 82-3, 101, 110-1, 115, 121
   Reynolds, William, 7, 9, 65, 131
   Smeaton, John, 24
   Snape, John, 10
   Stephenson, Robert, 33, 54
   Sword, J.A.S., 69, 73, 74
   Telford, Thomas, 7, 9, 53, 61-3, 65-6, 68, 116, 121, 122-3, 129, 131
   Trevithick, Richard, 80
   Upton, John, 116
   Weldon, Robert, 88
   Whitworth, Robert, 143, 145
   Woodhouse, John, 13
Essex, Earl of, 31

Failsworth, 101
Farleton Fell, *113*
Farnborough, 91
Fellows, Morton & Clayton, 17, 30, 52
Fitzwilliam, Earl, 134
Fleet, 91
Foxton, 53, 54, 122
Frampton-on-Severn, 17, 118

Gairlochy, 123
Garthmyl, 69
Gloucester, 116, 118-120, 143, *118*
Glyn Ceiriog, 68
Glyn Neath, 80
Gopsall Park, 147
Gosty Hill, 22-3, 17
Grand Union Canal Carrying Co., 30
Grantham, 139, 140
Great Glen, 121
Guilsfield, 69
Gumley, 52

Hadfield, Charles, 108
Halesowen, 22
Handel, George Frederick, 147
Harborough, Lord, 138
Harlaxton Hill, 141
Hebden Bridge, 103, 104
Heinz factory, 28
Hemel Hempstead, 32
Heptonstall, 104
Hereford, 143, 144

155

Hinckley,	145	Jura,	120
Hockley Heath,	18	Kendal,	110, 111
Homfray, Samuel,	80	Kensal Green,	28
Honeystreet,	84	Ketley,	7
Horseshoe Falls,	61, 68	King's Langley,	32
Horseley Ironworks,	22, 29, 44	Kintyre,	120
Huddersfield, 99,	105	Knighton,	17
Hughes, Ted,	104		
Hungerford,	85-6	Laggan,	123-4, *128*
Hutchings, David,	19	Lancaster,	110-1, *109*
		Landmark Trust,	15
Inclines		Lawrence, D.H.,	130
Combe Hay,	89	Leamington Spa,	48-9, *51*
Hay,	8, *23*	Ledbury,	144
Ketley,	8	Lee, Rev. William,	145
Trench,	9	Leicester,	52, 53, 56, *56*
Ironbridge,	9	Leighton Buzzard,	35
Ivinghoe,	46	Linslade,	35
		Little Eaton,	130, 131
James, Jack,	38	Little Venice,	27-8
Jacques, John,	147	Liverpool,	60
Junctions		Llangattock,	79
Aylestone,	56	Llangollen,	61, 62, 65
Barnsley,	133	Llanover,	78
Bordesley,	52	Llanymynach,	69, 71
Bull's Bridge,	28	Loch Dochfour,	121
Burgedin,	73	Loch Fyne,	120
Budbrooke,	49	Lochgilphead,	120
Castlefield,	99	Loch Linnhe,	121
Cosgrove,	38	Loch Lochy,	121, 122-3
Cowley Peachy,	30	Loch Ness,	121, 124
Ducie Street,	95, 101, 103	Loch Oich,	121, 123, 124
Dukinfield,	96, *102*	Locks	
Fairfield,	96	Ancoats,	95-6, *97*
Frankton,	65, 69, *66*	Ash,	90
Gayton,	42	Aston,	71
Hurlestone,	62, 69, *61*	Asylum,	29
King's Norton,	16-17, 18, *10*	Baddiley,	62
Kingswood,	18, 20, 52, *16*	Bascote,	48
Marston,	145	Belam,	74
Napton,	45, 46	Berkhamsted,	27
Norton,	53	Bingley,	122
Saul,	118, *118*	Bratch,	76
Swarkestone,	133	Brookwood,	89
Swinton,	133-4, *132*	Buckby,	44
Windmill End,	22, *21*	Caen Hill,	84, *87*

Index

Calcutt,	47	Bulbourne,	33, *36*
Cape,	49	Harby,	142
Carreghofa,	69, 71	Kilby,	*59*
Clachnaharry,	129	Mamilhad,	77
Corpach,	*123*	Manchester,	95, 99, 103, 110
Deepcut,	89-90, *92*	Market Bosworth,	146-7
Diggle,	*110*	Market Harborough,	53, 54, 56, *57*
Dunardry,	120	Market Overton,	139
Fort Augustus,	124, *127*	Marple,	98
Foxton,	53, 54-6, *57*	Marsden,	105-6, *106*
Grindley Brook,	63, 65, *63*	Marshall, Mrs.,	89
Hatton,	45, 50-1, *49, 50*	Measham,	148
Home Park Mill,	25	Melton Mowbray,	139
Knowle,	52	Merthyr Tydfil,	80
Kytra,	124, *126*	Midford,	86, 88
Lapworth,	20	Miller, Mr.,	143
Marple,	98	Milton Keynes,	37
Neptune's Staircase,	122, *124*	Moira,	145, 148, *147*
New Marton,	66	Montgomery,	69
Parkhead,	22	Morecambe Bay,	111
Rochdale summit,	*104*	Morriston,	81
Rogerstone,	76	Mumford, Sarah,	143
Rothersthorpe,	42, *54*	Myndd Llangatwg,	79
St. John's,	89		
Shelton,	133	Nantwich,	60, 62
Soulbury,	35, *35*	Nantyglo,	79
Springwell,	30	Napton,	46
Stockton,	*53*	Neath,	81
Stoke Bruerne,	*40*	Netherton,	22
Swanley,	62	New Bradwell,	37
Tardebigge,	12-13, 14-15, *13*	New Stanton,	130
Watford,	56	Newbury,	86
Welford Arm,	*58*	Newcomen engine,	134-6
Widcombe,	82	Newent,	144
Wilton,	44	Newnham,	94
Woodham,	*90*	Newport,	75
Woolsthorpe,	141	Newtown,	69, 70, 74
Wootton Rivers,	85	Nixon, Frank,	131
London, 17,	82	Northampton,	52
Long Itchington,	47	Norton,	52-3
		Nottingham,	142-3
McVities factory,	28	Nuneaton,	146
Madeley,	8, 61		
Maerdy,	73	Oaken Gates,	7
Maesbury Marsh,	71	Oakham,	138, 139
Maintenance yards,		Ochram Turn,	77

Odiham,	91-2, 94	Brecon & Abergavenny,	78-9
Old Oak Common,	28-9	Cromford & High Peak,	99
Ovaltine,	32	Grantham to Nottingham,	140
Oxford,	24	Great Western,	29, 76, 82
		Gronwen,	71
Paddington,	27, 26	Lancashire & Yorkshire,	103
Palmer, Graham,	70	London & Birmingham,	33, 54
Pant,	71	London, Midland & Scottish,	148
Paulton,	86	London & South Western,	89, 90
Peak District National Park,	98	Midland,	138
Peak Forest,	96	Monmouthshire,	75
Penarth,	74	Newent to Gloucester,	144
Pickford's,	41	Penydarren,	80
Pinkerton, John,	89	Shrewsbury & Chester,	70
Plas Kynaston,	68	Taff Vale,	75, 80
Plath, Sylvia,	104	West Highland,	121-2
Pontymoile,	77, 77	West Shropshire,	71
Pontypool,	75	Worcester to Birmingham,	15
Pontypridd,	80	Rednal,	70, 73
Pool Quay,	73	Reservoirs	
Porthywain,	69	Bittell,	15-16
Preston,	111, *111*	Blackstone Edge,	103
Pubs, hotels		Chelburn,	103
Blue Lias,	47	Cold Hendley,	138
Canal Tavern, Shrewsbury,	9	Daventry,	44
Cape of Good Hope,	49	Denton,	139, 141
Drawbridge,	19	Earlswood,	19
Hope and Anchor, Midford,	86	Hollingsworth,	103
Navigation, Aberdare,	80	Hurlestone,	62
Navigation, Maesbury,	71	Knipton,	139, 141
Plough, Hickling,	142	Lodge Farm,	22
Rutland Arms,	141	Mapperley,	130
Ship, Royston,	138	Marsworth,	33, *34*
Ship, Swinton,	134	Redbrook,	108
Towpath, Swinton,	134, *134*	Startop's,	33
Wharf, Hockley Heath,	19	Talybont,	79
		Tringford,	33
Quaker Oats,	29	Weston Turville,	33
Quaker's Yard,	80	Wilton Water,	85
Quina Brook,	65	Wintersett,	138
		Worsbrough,	136
Radstock,	86, 88	Resolven,	*81*
Railways and tramways		Rickmansworth,	30
Battlefield line,	146-7	Risca,	76
Birmingham & Stratford,	19	Rivers	
Birmingham to Gloucester,	14	Afon Lwyd,	76

Index

Avon,	16, 47, 49, 82	Vyrnwy,	72
Blyth,	52	Wey,	89
Blythe,	19	Witham,	139
Brent,	29	Wreak,	138
Blackwater,	90	Rogerstone,	81
Calder,	103, 104	Rolt, L.T.C.,	70
Ceiriog,	65, 66	Rowington,	51-2
Chess,	30, 31	Royston,	138
Cole,	19, 52	Ruabon,	62, 68
Colne,	30, 105	Rugby Cement,	47
Dearne,	136	Rutland, Duke of,	141-2
Dee,	60-1, 62, 65, 66, 68, 69		
Derwent,	130, 131	Saltisford,	49
Devon,	139	Salt Way,	15
Don,	133	Sandiacre,	130. 133
Douglas,	115	Seend Cleeve,	84, *86*
Erewash,	130	Selly Oak,	22
Eye,	138, 139	Semington,	83-4
Firth,	124	Severn & Canal Carrying Co.,	17
Gade,	30, 32	Shackerstone,	147
Goyt,	98	Shakespeare, William,	18, 19, 21
Great Ouse,	37, 38	Shardlow,	117
Itchen,	47	Shipley,	130
Kennet,	82, 85	Shrewley,	51
Leadon,	143	Shrewsbury,	62, 69
Leam,	48	Slaithwaite,	105, *107*
Lochy,	122	Southey, Robert,	122, 129
Lugg,	143	Sowerby Bridge,	99, 104
Mersey,	60, 68	Spondon,	131
Neath,	80	Stenwith,	141
Nene,	42-3, 52	Stockton,	47
Ness,	124	Stoke Bruerne,	38-40, 41, *39*
Ouzel,	35-7	Stoke Golding,	144-5
Perry,	70	Stourport,	10, 14, 117
Ribble,	115	Stratford-upon-Avon,	18, 20
Salwarpe,	15	Sutton Cheney,	146
Severn,	7, 8, 9, 10, 14, 15, 60, 62, 69, 72, 73, 74, 116, 117, 143, 144	Swansea,	60, 80, 81
		Swindon,	84
Soar,	52, 56	Swarkestone,	130
Tanat,	71	Sydney Gardens,	*84*
Tawe,	81		
Tern,	9, 131	Talybont,	79
Thames,	23, 24, 29, 33, 82, 84	Tardebigge lift,	13
Tove,	38	Tennant, George,	81
Trent,	56, 139, 143	Thimble Mill,	82
Usk,	75, 76, 79	Three Bridges,	29

159

Thrupp,	24	Watford Gap,	54
Ticknall,	144	Watling Street,	54
Todmorden,	103	Weedon,	43-4, *41*
Tring,	26, 32-3, 41	Welford,	54
Tunnels		Wellow,	88
Blisworth,	25-6, 37, 40-2, 44, *42*	Welsh Frankton,	62, 69, *66*
Brandwood,	19	Welshpool,	62, 70, 73, 74
Braunston,	25-6, 44	Wendover,	33
Brierley Hill,	8	Weston Lullingfields,	69
Broad Street, Birmingham,	18	Westwood,	83
Bruce,	85	Whaley Bridge,	99
Chirk,	68	Whitchurch,	63
Dudley,	22, *17*	Whixall Moss,	65
Edgbaston,	11-12, *17*	Wigan,	110
Gosty Hill,	22	Wolverhampton,	10
Greywell,	94	Wolverton,	37
Harecastle,	2, 9, 25, 106	Wombwell,	134
Hincastle,	111	Woodham,	89
Husbands Bosworth,	54	Worcester,	14
Hyde Bank,	96	Worcester Bar,	11, 14, 17-18, *11*
Kilsby,	54	Worsbrough,	133, 136
King's Norton, 10,	16	Wrenbury,	62, 63
Lapal,	22		
Oxenhall,	144	Yarranton, Andrew,	18
Queen Street, Cardiff,	80		
Saddington,	*55*		
Scout,	108		
Shortwood,	15		
Snarestone,	147		
Standedge,	99, 105, 106-8, *108*		
Shrewley,	51		
Shropshire Canal,	9		
Sydney Gardens, Bath,	82-3		
Tardebigge,	*12*		
Woodley,	96		
Tynley, Lord,	94		
Up Nately,	89		
Uppermill,	108		
Vyrnwy,	72, 73		
Waddington's,	133-4		
Wainlode,	143		
Warwick,	45, 47, 49		
Warwick Bar,	46, 52, 54		